SpringerBriefs in Probability and Mathematical Statistics

SpringerBriefs present concise summaries of cutting-edge research and practical applications across a wide spectrum of fields. Featuring compact volumes of 50 to 125 pages, the series covers a range of content from professional to academic. Briefs are characterized by fast, global electronic dissemination, standard publishing contracts, standardized manuscript preparation and formatting guidelines, and expedited production schedules.

Typical topics might include:

- A timely report of state-of-the art techniques
- A bridge between new research results, as published in journal articles, and a contextual literature review
- A snapshot of a hot or emerging topic
- Lecture of seminar notes making a specialist topic accessible for non-specialist readers
- SpringerBriefs in Probability and Mathematical Statistics showcase topics of current relevance in the field of probability and mathematical statistics

Manuscripts presenting new results in a classical field, new field, or an emerging topic, or bridges between new results and already published works, are encouraged. This series is intended for mathematicians and other scientists with interest in probability and mathematical statistics. All volumes published in this series undergo a thorough refereeing process.

The SBPMS series is published under the auspices of the Bernoulli Society for Mathematical Statistics and Probability.

More information about this series at http://www.springer.com/series/14353

Michael B. Marcus · Jay Rosen

Asymptotic Properties of Permanental Sequences

Related to Birth and Death Processes
and Autoregressive Gaussian Sequences

Bernoulli Society
for Mathematical Statistics
and Probability

Springer

Michael B. Marcus
253 West 73rd. St. Apt. 2E
New York, NY 10023
USA

Jay Rosen (ID)
Department of Mathematics
College of Staten Island
City University of New York
Staten Island, NY, USA

ISSN 2365-4333   ISSN 2365-4341 (electronic)
SpringerBriefs in Probability and Mathematical Statistics
ISBN 978-3-030-69484-5   ISBN 978-3-030-69485-2 (eBook)
https://doi.org/10.1007/978-3-030-69485-2

Mathematics Subject Classification: 60E07, 60G15, 60G17, 60G99, 60J27

This Springer imprint is published by the registered company Springer Nature Switzerland AG
The registered company address is: Gewerbestrasse 11, 6330 Cham, Switzerland

Michael Marcus dedicates this book to his dear friend the feminist literary scholar Jean Mills

Jay Rosen dedicates this book to his grandchildren

Preface

An R^n valued α-permanental random variable $(\tilde{X}_{\alpha,1}, \ldots, \tilde{X}_{\alpha,n})$ is a non-negative random variable with Laplace transform,

$$E\left(e^{-\sum_{i=1}^{n} s_i \tilde{X}_{\alpha,i}}\right) = \frac{1}{|I + \mathcal{K}\mathcal{S}|^\alpha}, \tag{1}$$

for some $n \times n$ matrix $\mathcal{K} = \mathcal{K}(i,j), i,j \in [1,n]$ and diagonal matrix $\mathcal{S}$ with positive entries $s_1, \ldots, s_n$, and $\alpha > 0$. We refer to $\mathcal{K}$ as the kernel of $(\tilde{X}_{\alpha,1}, \ldots, \tilde{X}_{\alpha,n})$.

An α-permanental process $X_\alpha = \{X_{\alpha,t}, t \in \mathcal{T}\}$ is a stochastic process that has finite-dimensional distributions which are α-permanental random variables. The permanental process is determined by a kernel $K = \{K(s,t), s,t \in \mathcal{T}\}$, with the property that for all $t_1, \ldots, t_n$ in $\mathcal{T}$, $\{K(t_i, t_j), i,j \in [1,n]\}$ determines the α-permanental random variable $(X_{\alpha,t_1}, \ldots, X_{\alpha,t_n})$ by (1). When (1) holds for a kernel K for all $\alpha > 0$, the family of permanental processes obtained are infinitely divisible. In this paper, we take $\mathcal{T} = \overline{\mathbb{N}}$, the strictly positive integers.

Eisenbaum and Kaspi show that the right hand side of (1) is the Laplace transform of a non-negative n-dimensional random variable for all $\alpha > 0$ if and only if $g\mathcal{K}g = \{g_i \mathcal{K}(i,j)g_j, i,j \in [1,n]\}$ is the potential of a transient symmetric Markov chain X_t with state space $[1,n]$ for some strictly positive sequence $\{g_i\}_{i\in[1,n]}$. That is,

$$g_i \mathcal{K}(i,j)g_j = E^i\left(\int_0^\infty 1_j(X_t)dt\right), \quad \forall i,j \in [1,n], \tag{2}$$

where E^i is the expectation operator for the Markov chain X_t starting at i, (i.e., $X_0 = i$). In this paper, we combine the $\{g_i\}_{i=1}^n$ with $\mathcal{K}$ and consider the matrix $\tilde{U} = g\mathcal{K}g$, which is the potential of a transient Markov chain. Similarly, we consider the kernel $\tilde{U} = \{\tilde{U}(s,t), s,t \in \mathcal{T}\}$ where,

$$\widetilde{U} = gKg, \tag{3}$$

which is the potential of a transient Markov chain on $\mathcal{T}$.

The fact that processes determined by Laplace transforms such as (1) are related to symmetric Markov chains is given by the Dynkin Isomorphism Theorem which is a relationship between stochastic processes that are the squares of Gaussian processes and the local times of symmetric Markov processes. These processes are linked in that the potential of the Markov chain is the covariance of the Gaussian process.

However, the kernel $\widetilde{U}$ is not necessarily symmetric. Significantly, Eisenbaum and Kaspi extended Dynkin's theorem to give a relationship between the local times of transient Markov processes with potential $\widetilde{U}$ that need not be symmetric and permanental processes with kernel $\widetilde{U}$. Therefore, one motivation for studying permanental processes is to obtain results about local times of transient Markov processes. Another motivation is that they are a challenging new class of positive infinitely divisible stochastic processes that require new techniques to analyze.

To be more specific, when $\widetilde{U}$ is symmetric, the $1/2$-permanental process it determines is the square of a Gaussian process, with covariance $\widetilde{U}$. But when $\widetilde{U}$ is not symmetric, the corresponding permanental process is not the square of a Gaussian process and cannot be analyzed using standard Gaussian process techniques. We obtain examples of kernels that are not symmetric by modifying symmetric kernels in the following way:

Let $U = \{U(j,k); j,k \in \overline{\mathbb{N}}\}$ be the potential of a symmetric transient Markov chain Y, and let $f = \{f(j); j \in \overline{\mathbb{N}}\}$ be an excessive function for Y. We consider kernels $\widetilde{U}$ of the form

$$\widetilde{U}(j,k) = U(j,k) + f(k), \quad j,k \in \overline{\mathbb{N}}. \tag{4}$$

We explain in Chap. 1 why $\widetilde{U}$ is the kernel of an α-permanental process and what are the excessive functions f.

We now briefly state the kinds of results obtained in this monograph. For a wide class of symmetric potentials U, the Gaussian process $\eta = \{\eta_j; j \in \overline{\mathbb{N}}\}$ with covariance U satisfies

$$\limsup_{j \to \infty} \frac{\eta_j}{(2\phi_j)^{1/2}} = 1 \quad a.s.,$$

for some sequence $\phi = \{\phi_j\}$. We show that this implies that for all $\alpha > 0$ the α-permanental sequence $\widetilde{X}_\alpha$, determined by the kernel $\widetilde{U}$ satisfies,

$$\limsup_{n\to\infty} \frac{\widetilde{X}_{\alpha,j}}{\phi_j} = 1 \quad a.s.,$$

when the excessive functions f satisfy $f_j = o(\phi_j)$, as $j \to \infty$.

The function ϕ is determined by U. Many examples are given in which U is the potential of symmetric birth and death processes with and without emigration and of Lévy processes on $\mathbf{Z}$. Also considered are Markov chains with potentials U that are the covariances of first and higher order autoregressive Gaussian sequences.

New York, USA   Michael B. Marcus
Staten Island, USA   Jay Rosen

Contents

1 **Introduction** 1
  1.1 General Results 5
  1.2 Applications 8

2 **Birth and Death Processes** 19

3 **Birth and Death Processes with Emigration** 33

4 **Birth and Death Processes with Emigration Related
  to First Order Gaussian Autoregressive Sequences** 47

5 **Markov Chains with Potentials That Are the Covariances
  of Higher Order Gaussian Autoregressive Sequences** 61

6 **Relating Permanental Sequences to Gaussian Sequences** 87

7 **Permanental Sequences with Kernels That Have Uniformly
  Bounded Row Sums** 97

8 **Uniform Markov Chains** 103

Bibliography 111

Index 113

Chapter 1
Introduction

We begin with a leisurely introduction to permanental processes and explain why they are interesting. Permanental processes are a recently discovered class of stochastic processes that, in some sense, generalize Gaussian processes. Whimsically, they can be thought of as Gaussian processes with covariances that are not necessarily symmetric. The background of how probabilists became interested in permanental processes is very interesting. It all started with a question raised by Paul Lévy which we now explain.

Let $\eta = (\eta_1, \ldots, \eta_n)$ be an n dimensional normal random variable with mean zero and covariance matrix

$$U = \{U_{j,k}\}_{j,k=1}^n. \tag{1.1}$$

Let $Z(1/2) = (Z_1, \ldots, Z_n) = (\eta_1^2/2, \ldots, \eta_n^2/2)$. It is well known and elementary that

$$E\left(e^{-\sum_{i=1}^n s_i Z_i}\right) = \frac{1}{|I + US|^{1/2}}, \tag{1.2}$$

where S is a diagonal matrix with entries $\{s_i; i \in [1, n]\}$. Now let

$$Z(k/2) = \sum_{i=1}^k Z_i(1/2), \tag{1.3}$$

where $Z_i(1/2) = (Z_{i,1}, \ldots, Z_{i,n}) = (\eta_{i,1}^2/2, \ldots, \eta_{i,n}^2/2), i = 1, \ldots, k$, are independent copies of $Z(1/2)$. Set $Z(k/2) = (Z_1(k/2), \ldots, Z_n(k/2))$. That is,

$$Z(k/2) = \left(\sum_{i=1}^k \frac{\eta_{i,1}^2}{2}, \ldots, \sum_{i=1}^k \frac{\eta_{i,n}^2}{2}\right). \tag{1.4}$$

© The Author(s), under exclusive license to Springer Nature Switzerland AG 2021
M. B. Marcus and J. Rosen, *Asymptotic Properties of Permanental Sequences*,
SpringerBriefs in Probability and Mathematical Statistics,
https://doi.org/10.1007/978-3-030-69485-2_1

It follows from (1.2) that

$$E\left(e^{-\sum_{i=1}^{n} s_i Z_i(k/2)}\right) = \frac{1}{|I + US|^{k/2}}.$$ (1.5)

We refer to $Z(k/2)$ as a chi-square random vector with k degrees of freedom.

Now suppose that

$$\frac{1}{|I + US|^{\alpha}}$$ (1.6)

is the Laplace transform of a random variable for all $\alpha > 0$. This implies that the random variable

$$\left(\frac{\eta_1^2}{2}, \ldots, \frac{\eta_n^2}{2}\right)$$ (1.7)

is infinitely divisible. These observations are probably what led Paul Lévy to ask the intriguing question, "When is a vector of Gaussian squares infinitely divisible?"

(Before we continue let us explain why we divide η^2 by 2. The answer is simple. We want U to be the covariance of η. If we considered η^2, the terms in the matrix U in (1.6) would have to be multiplied by 2.)

When (1.6) is the Laplace transform of a random variable for some $\alpha > 0$, we refer to U as its kernel and refer to the random variable as an α-permanental random variable. (This is because its joint moments can be expressed as α-permanents.) Having defined α-permanental random variables we can use the Kolmogorov Extension Theorem to define α-permanental sequences. We say $X_\alpha = \{X_{\alpha,j}; j \in \overline{\mathbb{N}}\}$, where $\overline{\mathbb{N}}$ denotes the strictly positive integers, is an α-permanental sequence with kernel

$$\mathcal{U} = \{U_{j,k}; j, k \in \overline{\mathbb{N}}\},$$ (1.8)

if for all n, $(X_{\alpha,1}, \ldots, X_{\alpha,n})$ is an n-dimensional α-permanental random variable with kernel

$$\mathcal{U}(n) := \{U_{j,k}; j, k \in [1, n]\}.$$ (1.9)

It is obvious that when (1.6) is the Laplace transform of a random variable for all $\alpha > 0$, the process X_α is infinitely divisible. So we come to the question: When is $\mathcal{U}$ the kernel of an infinitely divisible process? This is a rephrasing of the question originally posed by Lévy. Considering the way X_α is defined, the answer is whenever the Gaussian vector with covariance $\mathcal{U}(n)$ has infinitely divisible squares, for all possible n. Necessary and sufficient conditions for a Gaussian vector to have infinitely divisible squares are given in the papers of D. Vere-Jones [21], R. C. Griffiths [8] and R. Bapat [1]. However, what is important for us is the interpretation of their results by N. Eisenbaum and H. Kaspi. Eisenbaum and Kaspi, [5, Theorem 3.1] show that

$$\frac{1}{|I + \mathcal{U}(n)S|^{\alpha}}$$ (1.10)

is the Laplace transform of a non-negative n-dimensional random variable for all $\alpha > 0$ if and only if $g\mathcal{U}(n)g = \{g_i U_{i,j} g_j; i, j \in [1, n]\}$ is the potential of a transient symmetric Markov chain X_t with state space $[1, n]$ for some strictly positive sequence $\{g_i\}_{i \in [1,n]}$. That is, $g_i U_{i,j} g_j = E^i \left(\int_0^\infty 1_j (X_t) \, dt\right)$ for all $i, j \in [1, n]$, where E^i is the expectation operator for the Markov chain X_t starting at i, i.e. $X_0 = i$. In this paper we assume that we have made the choice of $\{g_i\}_{i \in [1,n]}$ and consider $\mathcal{U}(n)$ to be the potential of a transient Markov chain.

This holds in a more general setting. When $U = \{u(s, t), s, t \in \mathcal{T}\}$, is the potential density of a transient Markov process with state space $\mathcal{T}$, with respect to some σ-finite measure m on $\mathcal{T}$ and $u(s, t)$ is finite for all $s, t \in \mathcal{T}$, then U is the kernel of an α-permanental process X_α for all $\alpha > 0$. We refer to these permanental processes as associated α-permanental processes because they are associated with the transient Markov process.

It is astonishing that processes determined by Laplace transforms such as (1.10) should be related to Markov processes. In fact, the relationship is very deep. In [4] Dynkin obtains a remarkable result which shows that there is a relationship between stochastic processes that are the squares of Gaussian processes and the local times of symmetric Markov processes. These processes are linked in that the potential density of the Markov process is the covariance of the Gaussian process. In [10] we were able to use this relationship and results in the theory of Gaussian processes to give necessary and sufficient conditions for the joint continuity of the local times of strongly symmetric Markov processes.

Finally we come to the point where this becomes really interesting. The results of Eisenbaum and Kaspi do not require that the matrix in the Laplace transform is symmetric. Therefore in (1.10) the kernel $\mathcal{U}$ does not have to be symmetric and consequently the α-permanental sequences determined by these non-symmetric kernels are more general than those determined by the covariances of Gaussian sequences. Furthermore, Eisenbaum and Kaspi show that the Dynkin Isomorphism Theorem extends as well. So we now have a relationship between the local times of Markov chains that do not have symmetric potentials and α-permanental sequences. We can use this to obtain sample path properties of these local times. Clearly, the first step in this process is to find sample path properties of this more general class of permanental processes. In this monograph we do this for permanental processes with kernels that are modifications of the potentials of symmetric birth and death processes and auto regressive Gaussian sequences, but which themselves are not symmetric.

Even if permanental processes were not related to local times of Markov processes they still would be an interesting new class of processes to study because obtaining their sample path properties requires new techniques beyond those used to analyze Gaussian processes.

We proceed more formally repeating ourselves occasionally for clarity. An R^n valued α-permanental random variable $(\widetilde{X}_{\alpha,1}, \ldots, \widetilde{X}_{\alpha,n})$ is a non-negative random variable with Laplace transform

$$E \left(e^{-\sum_{i=1}^n s_i \widetilde{X}_{\alpha,i}}\right) = \frac{1}{|I + KS|^\alpha}, \tag{1.11}$$

for some $n \times n$ matrix K and diagonal matrix S with positive entries $s_1, \ldots, s_n$, and $\alpha > 0$. We refer to the matrix K as the kernel of $(\widetilde{X}_{\alpha,1}, \ldots, \widetilde{X}_{\alpha,n})$. An α-permanental process $\widetilde{X}_\alpha = \{\widetilde{X}_{\alpha,t}, t \in \mathcal{T}\}$ is a stochastic process that has finite dimensional distributions that are α-permanental random variables. In this paper we take $\mathcal{T} = \mathbb{N}$, the strictly positive integers, and refer to $\widetilde{X}_\alpha = \{\widetilde{X}_{\alpha,j}, j \in \mathbb{N}\}$ as an infinite dimensional α-permanental sequence.

Eisenbaum and Kaspi, [5, Theorem 3.1] show that the right hand side of (1.11) is the Laplace transform of a non-negative n-dimensional random variable for all $\alpha > 0$ if and only if $gKg = \{g_i K_{i,j} g_j, i, j \in [1, n]\}$ is the potential of a transient Markov chain with state space $[1, n]$ for some strictly positive sequence $\{g_i\}_{i=1}^n$. In this paper we combine the $\{g_i\}_{i=1}^n$ with K and consider $\widetilde{U} = gKg$, which is the potential of a transient Markov chain.

The matrix $\widetilde{U}$ is not necessarily symmetric. When it is not symmetric the corresponding permanental processes are really something new. We can obtain examples of kernels that are not symmetric by modifying symmetric kernels. Let X be a symmetric transient Markov chain with potential U and let $f = (f_1, \ldots)$ be an excessive function for X. We consider kernels $\widetilde{U}$ of the form

$$\widetilde{U}_{j,k} = U_{j,k} + f_k, \qquad j, k \in \overline{\mathbb{N}}. \tag{1.12}$$

Clearly, $\widetilde{U}$ is not symmetric. However, the kernels of α-permanental random variables are not unique. For example, if K satisfies (1.11) so does $\Lambda K \Lambda^{-1}$ for any $\Lambda \in \mathcal{D}_{n,+}$, the set of $n \times n$ diagonal matrices with strictly positive diagonal entries. We say that an $n \times n$ matrix K is equivalent to a symmetric matrix, or symmetrizable, if there exists an $n \times n$ symmetric matrix W such that,

$$|I + KS| = |I + WS| \quad \text{for all } S \in \mathcal{D}_{n,+}. \tag{1.13}$$

Nevertheless, it follows from [16, Theorem 1.1] that in Theorem 1.2 below we can always find excessive functions f such that $\{\widetilde{U}_{j,k}; l \leq j, k \leq n\}$ is not symmetrizable for all sufficiently large l and n. In fact we show in [16] that it is only in highly structured situations that the kernel of a permanental process is symmetrizable.

We are particularly interested in α-permanental processes that are not symmetrizable. Otherwise X is in some sense only a modification of a Gaussian process. This is not true when the kernel of α-permanental processes is not symmetrizable. In this case we get a new class of processes. These are the processes that we find particularly interesting.

The fact that $\widetilde{U}$ is the kernel of α-permanental processes is given by the next theorem, which is part of [15, Theorem 1.11].

Theorem 1.1 *Let $X = (\Omega, \mathcal{F}_t, X_t, \theta_t, P^x)$ be a symmetric transient Borel right process with state space $\mathbb{N}$, and strictly positive potential U. Then for any finite excessive function f for X and $\alpha > 0$, $\widetilde{U}$ is the kernel of an α-permanental sequence $\widetilde{X}_\alpha$.*

Recall that a non-negative function f is excessive for X, if $P_t f(x) \uparrow f(x)$ as $t \to 0$, for all x. The function f is a potential function of X if $f = Uh$ for some

$h \geq 0$. Since $Uh(x) = \int_0^\infty P_t h(x)\, dt$, it is easy to check that all potential functions are excessive. The potential functions that play a major role is this paper are $f = Uh$ where $h \in \ell_1^+$ or c_0^+. Note that since $U_{j,k} \leq U_{j,j} \wedge U_{k,k}$, (see [11, (13.2)]), when $h \in \ell_1^+$, $f_j = (Uh)_j < \infty$ for all $j \in \mathbb{N}$.

We use Theorem 1.1 to consider two families of α-permanental sequences; $\widetilde{X}_\alpha$ with kernel $\widetilde{U}$ and X_α with kernel U, and use the fact that $X_{1/2}$ is a sequence of Gaussian squares . The primary goal of this paper is to find sharp results about the asymptotic behavior of $\widetilde{X}_\alpha = \{\widetilde{X}_{\alpha,j}, j \in \mathbb{N}\}$ as $j \to \infty$. The way we proceed is find finite excessive functions f for X for which the asymptotic behavior of $\widetilde{X}_\alpha$ is the same as the asymptotic behavior of $X_{1/2}$. Obtaining the asymptotic behavior of $X_{1/2}$ is relatively simple because we are just dealing with Gaussian sequences. To be more explicit, we find finite excessive functions f such that

$$\limsup_{j\to\infty} \frac{\widetilde{X}_{\alpha,j}}{\phi_j} = \limsup_{j\to\infty} \frac{X_{1/2,j}}{\phi_j} \qquad a.s.. \tag{1.14}$$

The specific sequence of positive numbers $\phi = \{\phi_j\}$ is generally easily determined because $X_{1/2}$ is a sequence of Gaussian squares.

We get two classes of results. The first are general limit theorems for permanental processes that hold when their kernels U and $\widetilde{U}$ satisfy certain general conditions. These are Theorems 1.2–1.5 given in Sect. 1.1. In Sect. 1.2, in Theorems 1.6–1.11 we apply these results to the potentials of specific families of Markov chains. We consider birth and death processes, with and without emigration, and Markov chains with potentials that are the covariances of first and higher order autoregressive Gaussian sequences.

1.1 General Results

For any matrix K let $K(l, n)$ denote the $n \times n$ matrix obtained by restricting the matrix K to $\{l + 1, \ldots, l + n\} \times \{l + 1, \ldots, l + n\}$. In the next theorem we consider $U(l, n)^{-1}$. The reader should note that $(U(l, n))^{-1}$ is not generally the same as the matrix $U^{-1}(l, n)$.

For any invertible matrix M we often denote $M_{j,k}^{-1}$ by $M^{j,k}$.

Theorem 1.2 *Let X, U, f and $\widetilde{X}_\alpha$ be as in Theorem 1.1 and let η be a Gaussian sequence with covariance U. Then*

$$\sum_{k=1}^n (U(l, n))^{j,k} f_{k+l} \geq 0, \qquad j = 1, \ldots n. \tag{1.15}$$

Suppose, in addition that,

$$\sum_{j,k=1}^{n} (U(l,n))^{j,k} f_{k+l} = o_l(1), \quad \textit{uniformly in } n, \tag{1.16}$$

and there exists a sequence $\phi = \{\phi_j\}$ *such that,*

$$\limsup_{j\to\infty} \frac{\eta_j}{(2\phi_j)^{1/2}} = 1 \quad a.s., \tag{1.17}$$

and

$$f_j = o(\phi_j). \tag{1.18}$$

Then

$$\limsup_{j\to\infty} \frac{\widetilde{X}_{\alpha,j}}{\phi_j} = 1 \quad a.s. \tag{1.19}$$

for all $\alpha \geq 1/2$. *(Also, trivially, the upper bound holds for all* $\alpha > 0$.)

In most of our applications of this theorem we will use results in [15, Sect. 7] to show that (1.19) actually holds for all $\alpha > 0$.

The primary ingredient in Theorem 1.2 is the symmetric potential $U = \{U(j,k), j,k \in \bar{\mathbb{N}}\}$. We see in (1.16) that $(U(l,n))^{-1}$ must exist for all l and n. It follows from [11, Theorem 13.1.2] that this is the case.

Theorem 1.2 is proved in Chap. 6.

The next theorem gives limit theorems for permanental sequences $\widetilde{X}_\alpha$ when the row sums of U in (1.12) are uniformly bounded. It has a simpler more direct proof than Theorem 1.2 and doesn't require that we obtain the complicated estimate (1.16).

Theorem 1.3 *Let* X, U, f *and* $\widetilde{X}_\alpha$ *be as in Theorem 1.1. If*

$$\inf_j U_{j,j} > 0, \quad \sup_j \sum_{k=1}^{\infty} U_{j,k} < \infty, \quad \textit{and} \quad f \in c_0^+, \tag{1.20}$$

then

$$\limsup_{n\to\infty} \frac{\widetilde{X}_{\alpha,n}}{U_{n,n} \log n} = 1 \quad a.s. \tag{1.21}$$

Note that it follows from (1.20) that $\sup_n U_{n,n} < \infty$.

The proof of Theorem 1.2 uses a result that compares the permanental sequence $\widetilde{X}_\alpha$ with the Gaussian sequence η, determined by the covariance matrix U. Therefore U must be symmetric. The proof of Theorem 1.3 does not involve Gaussian processes and so we don't need U to be symmetric for that reason. The requirement that U must be symmetric is used because of Theorem 1.1. Theorem 6.1, [15] is similar to Theorem 1.1 but does not require that U is symmetric if f is a left potential function with respect to U, i.e., for all $k \in \bar{\mathbb{N}}$,

$$f_k = \sum_{j=1}^{\infty} h_j U_{j,k}, \quad \text{for some } h \in \ell_1^+. \tag{1.22}$$

See [15, (6.1)].

Using [15, Theorem 6.1] enables us to obtain limit theorems for permanental sequences with potential functions of the form of (1.12) in which U is the potential of Markov chains that are not necessarily symmetric.

Theorem 1.4 *Let $X = (\Omega, \mathcal{F}_t, X_t, \theta_t, P^x)$ be a transient Borel right process with state space $\overline{\mathbb{N}}$ and strictly positive potential U. Assume that*

$$\inf_j U_{j,j} > 0, \quad \sup_j \sum_{k=1}^{\infty} U_{j,k} < \infty, \quad \text{and} \quad \sup_k \sum_{j=1}^{\infty} U_{j,k} < \infty. \tag{1.23}$$

Let $f \in \ell_\infty$ be such that

$$f_k = \sum_{j=1}^{\infty} h_j U_{j,k}, \quad \text{for some } h \in \ell_1^+, \tag{1.24}$$

and let $\widetilde{U} = \{\widetilde{U}_{j,k}, j, k \in \overline{\mathbb{N}}\}$ where,

$$\widetilde{U}_{j,k} = U_{j,k} + f_k, \quad j, k \in \overline{\mathbb{N}}. \tag{1.25}$$

Then for any $\alpha > 0$, $\widetilde{U}$ is the kernel of an α-permanental sequence $\widetilde{X}_\alpha$ and

$$\limsup_{n \to \infty} \frac{\widetilde{X}_{\alpha,n}}{U_{n,n} \log n} = 1 \quad \text{a.s.} \tag{1.26}$$

Note that (1.25) looks the same as (1.12) but here U is not necessarily symmetric. Consequently, (1.26) is of interest even for $f = 0$. (See Example 8.1.)

Theorems 1.3 and 1.4 are proved in Chap. 7.

Let M be an $\overline{\mathbb{N}} \times \overline{\mathbb{N}}$ matrix and consider the operator norm on $\ell_\infty \to \ell_\infty$,

$$\|M\| = \sup_{\|x\|_\infty \le 1} \|Mx\|_\infty = \sup_j \sum_k |M_{j,k}|. \tag{1.27}$$

We say that a Markov chain X is uniform when its Q matrix has the property that $\|Q\| < \infty$. Since all the row sums of Q are negative,

$$\sup_j |Q_{j,j}| \le \|Q\| \le 2 \sup_j |Q_{j,j}|. \tag{1.28}$$

(For information on uniform Markov chains, see [7, Chap. 5].)

The next theorem allows us to replace the hypotheses of Theorem 1.3 with conditions on the Q matrix of X. Note that we call Q a $(2k + 1)$−diagonal matrix if $Q_{i,j} = 0$ for all $|j - i| > k$.

Theorem 1.5 *Let X, U, f and $\widetilde{X}_\alpha$ be as defined in Theorem 1.1 and assume furthermore that X is a uniform Markov chain. Then, if the row sums of the Q-matrix of X are bounded away from 0, and $f \in c_0^+$,*

$$\limsup_{n \to \infty} \frac{\widetilde{X}_{\alpha,n}}{U_{n,n} \log n} = 1 \qquad a.s. \tag{1.29}$$

Furthermore, when the Q-matrix is a $(2k + 1)$−diagonal matrix for some $k \geq 1$, $f \in c_0^+$ and $f = Uh$ for $h \in c_0^+$ are equivalent.

Theorem 1.5 is proved in Chap. 8.

1.2 Applications

The remaining theorems in this section, Theorems 1.6–1.11, are applications of the basic Theorems 1.2–1.5. The basic theorems give general results for the quadruple $(X, \widetilde{X}_\alpha, U, \widetilde{U})$. Our applications are examples based on specific choices of U. We use different symbols for the quadruple $(X, \widetilde{X}_\alpha, U, \widetilde{U})$ in the different examples.

The simplest examples of symmetric transient Markov chains are birth and death processes without emigration or explosion. We describe them by their Q matrix.

Let $s = \{s_j, \ j \geq 1\}$ be a strictly increasing sequence with $s_1 > 0$ and $\lim_{j \to \infty} s_j = \infty$, and let $Y = \{Y_t, t \in R^+\}$ be a continuous time birth and death process on $\overline{\mathbb{N}}$ with Q matrix $Q(s)$ where,

$$- Q(s) = \begin{pmatrix} a_1 + a_2 & -a_2 & 0 & \cdots & 0 & 0 & \cdots \\ -a_2 & a_2 + a_3 & -a_3 & \cdots & 0 & 0 & \cdots \\ \vdots & \vdots & \vdots & \ddots & \vdots & \vdots & \ddots \\ 0 & 0 & 0 & \cdots & a_{j-1} + a_j & -a_j & \cdots \\ 0 & 0 & 0 & \cdots & -a_j & a_j + a_{j+1} & \cdots \\ \vdots & \vdots & \vdots & \ddots & \vdots & \vdots & \ddots \end{pmatrix}, \tag{1.30}$$

and

$$a_1 = \frac{1}{s_1}, \quad \text{and} \quad a_j = \frac{1}{s_j - s_{j-1}}, \quad j \geq 1. \tag{1.31}$$

Since all the row sums are equal to 0, except for the first row sum, we see that Y is a birth and death process without emigration. (Except at the first stage. However, the first row can not also have a zero sum because if it did, the Markov process Y would not be transient.)

Since

$$s_j = \sum_{k=1}^{j} \frac{1}{a_k}, \tag{1.32}$$

the class of Q matrices in (1.30) include all symmetric birth and death processes for which

$$\sum_{k=1}^{\infty} \frac{1}{a_k} = \infty. \tag{1.33}$$

This implies that Y does not explode, that is, it does not run through all $\overline{\mathbb{N}}$ in finite time. See [20, Theorem 5.1].

We show in Theorem 2.1 that Y has potential $V = \{ V_{j,k}, \, j, k \in \overline{\mathbb{N}} \}$ where,

$$V_{j,k} = s_j \wedge s_k. \tag{1.34}$$

The next theorem is an application of Theorem 1.2 to the quadruple $(Y, \widetilde{Y}_\alpha, V, \widetilde{V})$. This is an example of $(X, \widetilde{X}_\alpha, U, \widetilde{U})$ in which $U = V$, in (1.34).

Let $s_i \uparrow \infty$ and define

$$\mathcal{K}_{\mathbf{s}}(j) = \log \left(\sum_{i=1}^{j-1} 1 \wedge \log(s_{i+1}/s_i) \right). \tag{1.35}$$

This function is introduced in [9] to obtain limit theorems for certain Gaussian sequences and is critical in our applications of Theorem 1.2.

Theorem 1.6 *Let V be as given in (1.34). Let $f = Vh$, where $h \in \ell_1^+$, (which implies that $f_j = g(s_j)$, $j \geq 1$, where g is an increasing strictly concave function) and let $\widetilde{Y}_\alpha = \{\widetilde{Y}_{\alpha,j}, \, j \in \overline{\mathbb{N}}\}$ be an α-permanental sequence with kernel $\widetilde{V} = \{\widetilde{V}_{j,k}; \, j, k \in \overline{\mathbb{N}}\}$, where*

$$\widetilde{V}_{j,k} = V_{j,k} + f_k, \qquad j, k \in \overline{\mathbb{N}}. \tag{1.36}$$

Then

$$\limsup_{j \to \infty} \frac{\widetilde{Y}_{\alpha,j}}{s_j \mathcal{K}_{\mathbf{s}}(j)} = 1, \qquad a.s., \qquad \forall \alpha > 0. \tag{1.37}$$

(We use the expression 'g is an increasing function' to include the case in which g is non-decreasing. We say that g is a strictly concave function when $\lim_{x \to \infty} g(x)/x = 0$.)

Properties of $\mathcal{K}_{\mathbf{s}}(j)$ are given in Lemma 2.7 and the examples following it. Using them we get the following corollary of Theorem 1.6.

Corollary 1.1 *In Theorem 1.6,*

(i) if $\limsup_{j \to \infty} s_j/s_{j-1} < \infty$, *then*

$$\limsup_{j\to\infty} \frac{\widetilde{Y}_{\alpha,j}}{s_j \log\log s_j} = 1, \quad a.s., \quad \forall \alpha > 0. \tag{1.38}$$

(ii) If $\liminf_{j\to\infty} s_j/s_{j-1} > 1$, *then*

$$\limsup_{j\to\infty} \frac{\widetilde{Y}_{\alpha,j}}{s_j \log j} = 1, \quad a.s., \quad \forall \alpha > 0. \tag{1.39}$$

We show in Chap. 2 that the potential functions $f = Vh$, where $h \in \ell_1^+$, satisfy (1.16) and (1.18). This allows us to apply Theorem 1.2. In Chap. 2 we also give a Riesz representation theorem for functions that are excessive for X.

In Chap. 3 we modify $Q(\mathbf{s})$, to obtain Q matrices for a large class of birth and death processes with emigration. Let $\mathcal{B} = \text{diag}(b_1, b_2, \ldots)$, i.e., $\mathcal{B}$ is a diagonal matrix with diagonal elements $(b_1, b_2, \ldots))$. Define

$$- \widetilde{Q}(\mathbf{s}) = -\widetilde{Q}(\mathbf{s}, \mathbf{b}) = \mathcal{B}(-Q(\mathbf{s}))\mathcal{B}. \tag{1.40}$$

We show that when $b_j = g(s_j)$, $j \geq 1$, where $g(x)$ is an increasing strictly concave function, $\widetilde{Q}(\mathbf{s})$ is the Q matrix of a birth and death process with emigration.

Let V be as given in (1.34) and let $W = \{W_{j,k}; j, k \in \overline{\mathbb{N}}\}$, where,

$$W_{j,k} = b_j^{-1} V_{j,k} b_k^{-1}, \quad j, k \geq 1. \tag{1.41}$$

We show in Lemma 3.2 that W is the potential of a Markov chain Z with Q matrix $\widetilde{Q}(\mathbf{s}, \mathbf{b})$.

The next theorem generalizes Theorem 1.6 and Corollary 1.1 .

Theorem 1.7 *Let W be as given in (1.41) and let $f = Wh$, where $h \in \ell_1^+$. Let $\widetilde{Z}_\alpha = \{\widetilde{Z}_{\alpha,j}, j \in \overline{\mathbb{N}}\}$ be an α-permanental sequence with kernel $\widetilde{W} = \{\widetilde{W}_{j,k}; j, k \in \overline{\mathbb{N}}\}$, where*

$$\widetilde{W}_{j,k} = W_{j,k} + f_k, \quad j, k \in \overline{\mathbb{N}}. \tag{1.42}$$

Then

$$\limsup_{j\to\infty} \frac{\widetilde{Z}_{\alpha,j}}{W_{j,j}\mathcal{K}_\mathbf{s}(j)} = 1, \quad a.s., \quad \forall \alpha > 0. \tag{1.43}$$

If $\limsup_{j\to\infty} s_j/s_{j-1} < \infty$, *then*

$$\limsup_{j\to\infty} \frac{\widetilde{Z}_{\alpha,j}}{W_{j,j} \log\log s_j} = 1, \quad a.s., \quad \forall \alpha > 0. \tag{1.44}$$

If $\liminf_{j\to\infty} s_j/s_{j-1} > 1$, *then*

$$\limsup_{j \to \infty} \frac{\widetilde{Z}_{\alpha,j}}{W_{j,j} \log j} = 1, \qquad a.s., \qquad \forall \alpha > 0. \tag{1.45}$$

In Lemma 3.4 we show under the hypotheses of Theorem 1.7, $\lim_{j\to\infty} f_j/W_{j,j} = 0$.

Clearly, when $\mathcal{B}$ is the identity matrix, Theorem 1.7 gives Theorem 1.6, It is useful to state Theorem 1.6 separately because it is instrumental in the proof of Theorem 1.7.

An interesting class of examples of potentials of the form of (1.41) is when $b_j = s_j^{1/2}$ for all $j \in \overline{\mathbb{N}}$. We denote this potential by $\mathcal{W} = \{W_{j,k}, j, k \in \overline{\mathbb{N}}\}$, and see that

$$W_{j,k} = \frac{b_j}{b_k}, \qquad j \le k. \tag{1.46}$$

This expression is much more interesting if we set $b_j = e^{v_j}$. Then we get

$$W_{j,k} = e^{-|v_k - v_j|}, \qquad \forall j, k \in \overline{\mathbb{N}}. \tag{1.47}$$

Let $\xi = \{\xi(x), x \in R^+\}$ be a mean zero Gaussian process with covariance $\exp(-|x - y|)$ and note that ξ is an Ornstein-Uhlenbeck process and $\mathcal{W}$ is the covariance of $\{\xi(v_j), j \in \overline{\mathbb{N}}\}$. In Theorem 3.1 we show what results Theorem 1.7 give when W is written as in (1.47).

In Chap. 4 we take the potentials, described abstractly in (1.41), to be the covariance of a first order auto regressive Gaussian sequence. Let $\{g_j, j \in \overline{\mathbb{N}}\}$ be a sequence of independent identically distributed standard normal random variables and $\{x_n\}$ an increasing sequence with $0 < x_n \le 1$. We consider first order autoregressive Gaussian sequences $\widehat{\xi} = \{\widehat{\xi}_n, n \in \overline{\mathbb{N}}\}$, defined by,

$$\widehat{\xi}_1 = g_1, \qquad \widehat{\xi}_n = x_{n-1}\widehat{\xi}_{n-1} + g_n, \qquad n \ge 2. \tag{1.48}$$

The covariance of $\widehat{\xi}$ is $\mathcal{U} = \{\mathcal{U}_{j,k}, j, k \in \overline{\mathbb{N}}\}$, where,

$$\mathcal{U}_{j,k} = \sum_{i=1}^{j} \left(\prod_{l=i}^{j-1} x_l \prod_{l=i}^{k-1} x_l \right), \qquad j \le k. \tag{1.49}$$

and $\{x_j\}$ is an increasing sequence, with $0 < x_j \le 1$. This has the form of (1.41) with

$$b_j = \prod_{l=1}^{j-1} x_l^{-1}, \qquad \text{and} \qquad s_j = \mathcal{U}_{j,j}b_j^2 = \sum_{i=1}^{j} b_j^2, \tag{1.50}$$

and consequently, as we show, is the potential of a Markov chain which we denote by $\mathcal{X}$. In addition we show in Lemma 4.3 that $\lim_{j\to\infty} \mathcal{U}_{j,j}$ exists and is strictly greater than 1. In this case Theorem 1.7 gives:

Theorem 1.8 *Let $\mathcal{U}$ be as given in (1.49) and let f be a finite excessive function for X. Let $\tilde{X}_\alpha = \{\tilde{X}_{\alpha,j}, j \in \bar{\mathbb{N}}\}$ be an α-permanental sequence with kernel $\tilde{\mathcal{U}} = \{\tilde{\mathcal{U}}_{j,k}; j, k \in \bar{\mathbb{N}}\}$, where*

$$\tilde{\mathcal{U}}_{j,k} = \mathcal{U}_{j,k} + f_k, \qquad j, k \in \bar{\mathbb{N}}. \tag{1.51}$$

(i) *If $\lim_{j\to\infty} \mathcal{U}_{j,j} = \infty$, or equivalently, $\lim_{j\to\infty} x_j = 1$, and $f = \mathcal{U}h$, where $h \in \ell_1^+$, then*

$$\limsup_{j\to\infty} \frac{\tilde{X}_{\alpha,j}}{\mathcal{U}_{j,j} \log\log(\mathcal{U}_{j,j}b_j^2)} = 1 \quad a.s. \tag{1.52}$$

In particular, if $\mathcal{U}_{j,j}$ is a regularly varying function with index $0 < \beta < 1$, then

$$\limsup_{j\to\infty} \frac{\tilde{X}_{\alpha,j}}{\mathcal{U}_{j,j} \log j} = 1 - \beta \quad a.s. \tag{1.53}$$

(ii) *If $\lim_{j\to\infty} \mathcal{U}_{j,j} = 1/(1 - \delta^2)$, for some $0 < \delta < 1$, or equivalently, $\lim_{j\to\infty} x_j = \delta < 1$, and $f \in c_0^+$, then*

$$\limsup_{j\to\infty} \frac{\tilde{X}_{\alpha,j}}{\log j} = \frac{1}{1 - \delta^2} \quad a.s. \tag{1.54}$$

Furthermore, when $\lim_{j\to\infty} \mathcal{U}_{j,j} = 1/(1 - \delta^2)$, for some $0 < \delta < 1$, $f \in c_0^+$ and $f = \mathcal{U}h$ for $h \in c_0^+$ are equivalent.

The statement in (1.52) and even the one in (1.53) do not seem too useful because there are too many unknowns. Ultimately everything depends on the sequence $\{x_j\}$. We give some examples. They are arranged in order of decreasing values of x_j, (for large j).

Example 1.1 (i) If $j(1 - x_j^2) \to 0$ as $j \to \infty$,

$$\limsup_{j\to\infty} \frac{\tilde{X}_{\alpha,j}}{j \log\log j} = 1 \quad a.s., \qquad \forall \alpha > 0. \tag{1.55}$$

This includes the case where $\prod_{j=0}^\infty x_j > 0$.

(ii) If $j(1 - x_j^2) \sim c$ as $j \to \infty$ for some $c > 0$,

$$\limsup_{j\to\infty} \frac{\tilde{X}_{\alpha,j}}{j \log\log j} = \frac{1}{1+c}, \quad a.s., \qquad \forall \alpha > 0. \tag{1.56}$$

(iii) If $j^\beta(1 - x_j^2) \sim 1$ as $j \to \infty$, for $0 < \beta < 1$,

$$\limsup_{j \to \infty} \frac{\widetilde{\mathcal{X}}_{\alpha,j}}{j^\beta \log j} = 1 - \beta, \quad a.s., \qquad \forall \alpha > 0. \tag{1.57}$$

In Chap. 5 we take the symmetric potential U in (1.12) to be the covariance of a k-th order autoregressive Gaussian sequence, $k \geq 2$. Let $\{g_j, j \in \overline{\mathbb{N}}\}$ be a sequence of independent identically distributed standard normal random variables and $\{p_i\}_{i=1}^k$ a decreasing sequence of probabilities with $\sum_{l=1}^k p_l \leq 1$. We define the Gaussian sequence $\xi = \{\xi_n, n \in \overline{\mathbb{N}}\}$ by,

$$\xi_1 = g_1, \quad \text{and} \quad \xi_n = \sum_{l=1}^k p_l \xi_{n-l} + g_n, \qquad n \geq 2, \tag{1.58}$$

where $\xi_i = 0$ for all $i \leq 0$. Let

$$V = \{V_{m,n}, m, n \in \overline{\mathbb{N}}\}, \tag{1.59}$$

be the covariance of ξ.

We show that with certain additional conditions, V is the potential of a continuous time Markov chain $\mathcal{Y}$ on $\overline{\mathbb{N}}$ with a Q matrix that is a symmetric Töeplitz matrix which is completely determined by $\{Q_{n,m}\}_{m \geq n}$, i.e.,

$$Q_{n,n} = -\left(1 + \sum_{l=1}^k p_l^2\right), \quad Q_{n,j} = \beta_j > 0, \quad j \in [n+1, n+k],$$
$$Q_{n,j} = 0, \quad j > n + k, \tag{1.60}$$

where β_j are functions of $\{p_i\}_{i=1}^k$. In addition, the row sums of the n-th row of Q, for $n \geq k + 1$, is equal to $(1 - \sum_{l=1}^k p_l)^2$.

We can consider these Markov chains as population models which, when at stage $n \geq k + 1$, increase or decrease by 1 to k members, and so are generalizations of birth and death processes. When $\sum_{l=1}^k p_l = 1$, there is no emigration once the population size reaches k. When $\sum_{l=1}^k p_l < 1$, there is emigration at each stage.

Theorem 1.9 *Let V be as defined in (1.59) with the additional property that $p_i \downarrow$, and let f be a finite excessive function for $\mathcal{Y}$. Let $\widetilde{\mathcal{Y}}_\alpha = \{\widetilde{\mathcal{Y}}_{\alpha,j}, j \in \overline{\mathbb{N}}\}$ be an α-permanental sequence with kernel $\widetilde{V} = \{\widetilde{V}_{j,k}; j, k \in \overline{\mathbb{N}}\}$ where,*

$$\widetilde{V}_{j,k} = V_{j,k} + f_k. \tag{1.61}$$

(i) If $\sum_{j=1}^k p_j < 1$ and $f \in c_0^+$, then

$$\limsup_{j \to \infty} \frac{\widetilde{\mathcal{Y}}_{\alpha,j}}{\log j} = c^*, \qquad \forall \alpha > 0, \qquad a.s. \tag{1.62}$$

for some constant

$$1 + p_1^2 \le c^* \le \frac{1}{1 - (\sum_{l=1}^{k} p_l)^2}. \tag{1.63}$$

The precise value of c^* is given in (5.135).

(ii) *If $\sum_{j=1}^{k} p_j = 1$ and in addition $f = Vh$ where $h \in \ell_1^+$, then*

$$\limsup_{j \to \infty} \frac{\tilde{y}_{\alpha,j}}{j \log \log j} = \frac{1}{\left(\sum_{l=1}^{k} l p_l \right)^2} \qquad a.s., \qquad \forall \alpha \ge 1/2. \tag{1.64}$$

Furthermore, when $\sum_{j=1}^{k} p_j < 1$, $f \in c_0^+$ and $f = Vh$ for $h \in c_0^+$ are equivalent.

The limits in (1.64) and (1.62) may also hold for certain sequences $\{p_i\}$ that are not decreasing. See Remark 5.1.

We show in Lemma 5.13 that when $\sum_{j=1}^{k} p_j = 1$, the condition $f = Vh$ where $h \in \ell_1^+$, holds for all concave increasing functions f satisfying $f_j = o(j)$ as $j \to \infty$. Furthermore it is trivial that the upper bound in (1.64) holds for all $\alpha > 0$. But we need additional conditions on the potential functions f to show that the lower bound holds for all $\alpha > 0$.

In the next theorem we show that (1.64) holds for all $\alpha > 0$, when the potential functions f are such that $f_j = o(j^{1/2})$ as $j \to \infty$. We don't think that this restriction is required but we need it to use the techniques that we have at our disposal.

Theorem 1.10 *Under the hypotheses of Theorem 1.9 assume in addition that $f_j = o(j^{1/2})$ as $j \to \infty$. Then (1.64) holds for all $\alpha > 0$.*

When $\sum_{j=1}^{k} p_j < 1$, the condition $f = Vh$ where $h \in \ell_1^+$, implies that $f \in \ell_1$. In Remark 5.2 we give an explicit formula for c^* in terms of the roots of the polynomial

$$P(x) = 1 - \left(\sum_{l=1}^{k} p_l x^l \right). \tag{1.65}$$

We use ideas from the proofs of Theorems 1.3 and 1.4 to get limit theorems for permanental sequences with kernels that are related to the potentials of Lévy processes that are not necessarily symmetric.

Theorem 1.11 *Let X be a Lévy process on $\mathbb{Z}$ that is killed at the end of an independent exponential time, with potential $U = \{U_{j,k}; \ j, k \in \mathbb{Z}\}$.*

Let $f = \{f_k, k \in \mathbb{Z}\}$ be a finite excessive function for X, and let $\tilde{U} = \{\tilde{U}_{j,k}, \ j, k \in \mathbb{Z}\}$ where,

$$\tilde{U}_{j,k} = U_{j,k} + f_{-k}, \qquad j, k \in \mathbb{Z}. \tag{1.66}$$

Then for any $\alpha > 0$, $\tilde{U}$ is the kernel of an α-permanental sequence $\tilde{X}_\alpha$, and if $\lim_{k \to \infty} f_{-k} = 0$, then

$$\limsup_{n \to \infty} \frac{\widetilde{X}_{\alpha,n}}{\log n} = U_{0,0}. \qquad a.s., \tag{1.67}$$

Furthermore, if $g = Uh$, for a positive sequence h, then $g \in c_0^+(\mathbb{Z})$ if and only if $h \in c_0^+(\mathbb{Z})$.

Note that when X is not symmetric, (1.67) is of interest even for $f = 0$.

Theorems 1.1–1.5, are results for the broad classes of permanental processes described by quadruple $(X, \widetilde{X}_\alpha, U, \widetilde{U})$. Theorem 1.1 is given in [15, Theorem 1.11]. Theorem 1.2 is proved in Chap. 6. Theorem 1.3 and 1.4 are proved in Chap. 7 and Theorem 1.5 is proved in Chap. 8. Theorems 1.6–1.11 are applications of Theorems 1.1–1.5 in which the matrices U are the potential of specific families of Markov chains. We use different symbols for $(X, \widetilde{X}_\alpha, U, \widetilde{U})$ in the different examples.

In Chap. 2 we take $U = V = \{V_{j,k}, j, k \in \overline{\mathbb{N}}\}$ where,

$$V_{j,k} = s_j \wedge s_k, \qquad \text{for } s_j \uparrow \infty, \tag{1.68}$$

and give the proof of Theorem 1.6.

In Chap. 3 we take $U = W = \{W_{j,k}, j, k \in \overline{\mathbb{N}}\}$ where,

$$W_{j,k} = \frac{s_j \wedge s_k}{b_j b_k} \tag{1.69}$$

and $b = \{b_j\}$ is a finite potential function for the Markov process determined by V. Theorem 1.7 is proved in this section. We consider the specific example given in (1.46) and (1.47) in which, $U = \mathcal{W} = \{\mathcal{W}_{j,k}, j, k \in \overline{\mathbb{N}}\}$ where,

$$\mathcal{W}_{j,k} = e^{-|v_k - v_j|}, \qquad \forall\, j, k \in \overline{\mathbb{N}}, \tag{1.70}$$

For a sequence $v_j \uparrow \infty$. Theorem 3.1 gives limit theorems for permanental processes based $\mathcal{W}$.

In Chap. 4 we take $U = V$ to be the covariance of a first order autoregressive Gaussian sequence. In this case $\mathcal{U}_{j,k}$ is also an example of (1.69) in which

$$b_j = \prod_{l=1}^{j-1} x_l^{-1}, \qquad \text{and} \qquad s_j = \sum_{i=1}^{j} b_i^2. \tag{1.71}$$

We give the proof of Theorem 1.8 in this section.

In Chaps. 2–4 the potentials are all examples of (1.69). The Markov chains with these potentials only move between their nearest neighbors. In Chap. 5 we take the symmetric potential U in (1.12) to be the covariance of a k-th order autoregressive Gaussian sequences for $k \geq 2$, and denote it by V. Markov chains with these potentials move amongst their k nearest neighbors. We can not find the potentials of these chains precisely but we can estimate the potentials sufficiently well to give a proof of Theorems 1.9 and Theorem 1.10.

Remark 1.1 Lévy's question, "When is a vector of Gaussian squares infinitely divisible?", was reintroduced by Vere-Jones [21] and answered by Giffiths [8] and Bapat [1]. This work led the way to the consideration of permanental processes with symmetric kernels. We discuss this in great detail in [11, Sect. 13.2]. The infinite divisibility of a vector of Gaussian squares is not really relevant to the work in this monograph. For one thing, we are concerned with permanental processes with kernels that are not symmetric. Nevertheless, given Lévy's prominence, it is interesting to speculate about what he had in mind. It turns out that even for two dimensional Gaussian random variables say (η_1, η_2) this is a difficult question and except when η_1 and η_2 are independent we can not find a simple example in which the vector (η_1^2, η_2^2) is infinitely divisible. In fact Lévy conjectured that not all two dimensional Gaussian vectors are infinitely divisible, which is false. In the following lemma, using elementary ideas, we show that there are many two dimensional vectors of Gaussian squares that are infinitely divisible. We also see, given the condition we must impose, why one might guess that this is not true for all two dimensional vectors of Gaussian squares.

Lemma 1.1 *Let (η_1, η_2) be a Gaussian random variable with covariance*

$$U = \frac{1}{ab - c^2} \begin{pmatrix} b & c \\ c & a \end{pmatrix}, \tag{1.72}$$

where $a.b, c > 0$ and,

$$a^2 \wedge b^2 > ab - c^2. \tag{1.73}$$

Then

$$\left(\frac{\eta_1^2}{2}, \frac{\eta_2^2}{2} \right) \tag{1.74}$$

is infinitely divisible.

Proof Let

$$A = U^{-1} = \begin{pmatrix} a & -c \\ -c & b \end{pmatrix}, \tag{1.75}$$

and note that

$$|A||I + US| = |A + S|, \tag{1.76}$$

so that,

$$\frac{1}{|I + US|^\alpha} = \frac{|A|^\alpha}{|A + S|^\alpha} := F(s_1, s_2). \tag{1.77}$$

We now show that when (1.73) holds $F(s_1, s_2)$ is completely monotone for s_1, s_2 in the neighborhood of the origin and consequently is a Laplace transform. We have,

$$F(s_1, s_2) = \frac{|A|^\alpha}{|(s_1 + a)(s_2 + b) - c^2|^\alpha}. \tag{1.78}$$

Therefore,

$$\frac{\partial F(s_1, s_2)}{\partial s_1} = -\alpha \frac{(s_2 + b)|A|^\alpha}{|(s_1 + a)(s_2 + b) - c^2|^{\alpha+1}}, \tag{1.79}$$

and similarly for the derivative with respect to s_2. In either case the derivative changes sign. Now consider,

$$\frac{\partial^n F(s_1, s_2)}{\partial^l s_1 \partial^{n-l} s_2} = \sum C(p, j_p, k_p) \frac{(s_1 + a)^{j_p}(s_2 + b)^{k_p}}{|(s_1 + a)(s_2 + b) - c^2|^{\alpha+p}}, \tag{1.80}$$

where $C(p, j_p, k_p)$ is a real number, which may be 0, and the sum is taken over $[n/2] \leq p \leq n$ and $0 \leq j_p + k_p \leq p$. We have,

$$\frac{\partial}{\partial s_1} \frac{(s_1 + a)^{j_p}(s_2 + b)^{k_p}}{|(s_1 + a)(s_2 + b) - c^2|^{\alpha+p}} \tag{1.81}$$

$$= \frac{(s_1 + a)^{j_p-1}(s_2 + b)^{k_p}}{|(s_1 + a)(s_2 + b) - c^2|^{\alpha+p}} \left(\frac{-(\alpha + p)(s_1 + a)^2}{(s_1 + a)(s_2 + b) - c^2} + j_p \right).$$

By (1.73), $a^2 > ab - c^2$. Therefore, for all s_1, s_2 sufficiently small,

$$(s_1 + a)^2 \geq (s_1 + a)(s_2 + b) - c^2, \tag{1.82}$$

which implies that,

$$(\alpha + p)(s_1 + a)^2 \geq j_p \left((s_1 + a)(s_2 + b) - c^2 \right). \tag{1.83}$$

This shows that the derivative in (1.81) is negative. Therefore, taking the derivative in (1.80) with respect to s_1 changes the sign of the sum. A similar analysis shows that taking the derivative in (1.80) with respect to s_2 also changes the sign of the sum. Therefore $F(s_1, s_2)$ is completely monotone in the neighborhood of the origin. □

We thank Pat Fitzsimmons and Kevin O'Bryant for several helpful conversations.

Chapter 2
Birth and Death Processes

Let $\mathbf{s} = \{s_j, \ j \geq 1\}$ be a strictly increasing sequence with $s_j > 0$ and $\lim_{j \to \infty} s_j = \infty$, and let $\overline{Y} = \{\overline{Y}_t, t \in R^+\}$ be the continuous time birth and death process on $\overline{\mathbb{N}}$, without emigration, with Q matrix $\overline{Q}(\mathbf{s})$ where,

$$
-\overline{Q}(\mathbf{s}) = \frac{1}{2}
\begin{pmatrix}
a_1 + a_2 & -a_2 & 0 & \cdots & 0 & 0 & \cdots \\
-a_2 & a_2 + a_3 & -a_3 & \cdots & 0 & 0 & \cdots \\
\vdots & \vdots & \vdots & \ddots & \vdots & \vdots & \ddots \\
0 & 0 & 0 & \cdots & a_{j-1} + a_j & -a_j & \cdots \\
0 & 0 & 0 & \cdots & -a_j & a_j + a_{j+1} & \cdots \\
\vdots & \vdots & \vdots & \ddots & \vdots & \vdots & \ddots
\end{pmatrix},
\tag{2.1}
$$

and

$$
a_1 = \frac{1}{s_1}, \quad \text{and} \quad a_j = \frac{1}{s_j - s_{j-1}}, \quad j \geq 1.
\tag{2.2}
$$

Since

$$
s_j = \sum_{k=1}^{j} \frac{1}{a_k},
\tag{2.3}
$$

the class of Q matrices in (2.1) include all symmetric birth and death processes for which

$$
\sum_{k=1}^{\infty} 1/a_k = \infty.
\tag{2.4}
$$

This implies that $\{\overline{Y}_t, t \in R^+\}$ does not explode, that is it, does not run through all $\overline{\mathbb{N}}$ in finite time; see [20, Theorem 5.1].

© The Author(s), under exclusive license to Springer Nature Switzerland AG 2021
M. B. Marcus and J. Rosen, *Asymptotic Properties of Permanental Sequences*,
SpringerBriefs in Probability and Mathematical Statistics,
https://doi.org/10.1007/978-3-030-69485-2_2

Theorem 2.1 *The continuous time birth and death process $\overline{Y}$ has potential*

$$\overline{V}_{j,k} = 2\left(s_j \wedge s_k\right), \quad j, k \in \overline{\mathbb{N}}. \tag{2.5}$$

Proof It is easy to see that $\overline{V}\,\overline{Q}(s) = \overline{Q}(s)\overline{V} = -I$ in the sense of matrix multiplication. However, generally, this is not sufficient to show that $\overline{Y}$ has potential $\overline{V}_{j,k}$,
(unless $\sup_j a_j < \infty$, see Lemma 5.4). We see in Lemma 2.2 that there are functions
f with $\overline{Q}(s)f = 0$.

Let $\overline{B} = \{\overline{B}_t, t \in R^+\}$ be Brownian motion killed the first time it hits 0. $\overline{B}$ has
potential densities

$$U_{\overline{B}}(x, y) = 2\left(x \wedge y\right), \quad x, y > 0. \tag{2.6}$$

We use $\overline{B}$ to prove (2.5). To do this we first make the connection between $\overline{Y}$ and $\overline{B}$.

Using (2.1) and the relationship between the **Q** matrix and the jump matrix of
the Markov chain, (see [17, Chap. 2.6]), we have that for all $n \geq 2$,

$$P_{\overline{Y}}(n, n+1) = \frac{a_{n+1}}{a_n + a_{n+1}} \tag{2.7}$$

$$= \frac{s_n - s_{n-1}}{s_{n+1} - s_{n-1}} = P_{\overline{B}}^{s_n}\left(T_{s_{n+1}} < T_{s_{n-1}}\right),$$

where we use [18, Chap. II, Proposition 3.8] for the last equality. (As usual, T_x is the
first hitting time of x.)

Similarly,

$$P_{\overline{Y}}(n, n-1) = \frac{a_n}{a_n + a_{n+1}} \tag{2.8}$$

$$= \frac{s_{n+1} - s_n}{s_{n+1} - s_{n-1}} = P_{\overline{B}}^{s_n}\left(T_{s_{n-1}} < T_{s_{n+1}}\right).$$

In the same manner we have,

$$P_{\overline{Y}}(1, 2) = \frac{a_2}{a_1 + a_2} = P_{\overline{B}}^{s_1}\left(T_{s_2} < T_\Delta\right), \tag{2.9}$$

where Δ is the cemetery state, and

$$P_{\overline{Y}}(1, \Delta) = \frac{a_1}{a_1 + a_2} = P_{\overline{B}}^{s_1}\left(T_\Delta < T_{s_2}\right). \tag{2.10}$$

Now, let L_t^x denote the local time of Brownian motion. It follows from [18, Chap.
VI, (2.8)], that for all $n \geq 1$,

$$E_{\overline{B}}^{s_n}\left(L_{T_{s_{n-1}}\wedge T_{s_{n+1}}}^{s_n}\right) = 2\frac{(s_{n+1} - s_n)\,(s_n - s_{n-1})}{s_{n+1} - s_{n-1}} \tag{2.11}$$

$$= \frac{2}{a_n + a_{n+1}}.$$

We see from [17, Sect. 2.6] and the Q matrix in (2.1) that the holding time of $\overline{Y}$ at n is an exponential random variable with parameter $(a_n + a_{n+1})/2$ that is independent of everything else. This holding time has expectation $2/(a_n + a_{n+1})$.

To obtain (2.5) we show that the behavior of $\overline{Y}$ and $\overline{B}$ are similar in the following sense: Begin $\overline{Y}$ at j and $\overline{B}$ at s_j. The next visit of $\overline{Y}$ to an integer will be to either $j + 1$, with probability (2.7), or to $j - 1$ with probability (2.8). These are the same probabilities that the next visit of $\overline{B}$ is to s_{j+1} or s_{j-1}. During the time interval that $\overline{Y}$ and $\overline{B}$ make this transition, it follows from the last paragraph that the expected value of the increase in $L_t^{s_j}$ is the expected amount of time that $\overline{Y}$ spends at j. We repeat this analysis until the processes move to Δ, at which time they die. It follows from this that,

$$\overline{V}_{j,k} = E^j\left(\int_0^\infty 1_k\left(\overline{Y}_t\right)\,dt\right) = E_{\overline{B}}^{s_j}\left(L_\infty^{s_k}\right) = U_{\overline{B}}(s_j, s_k), \tag{2.12}$$

which, by (2.6), gives (2.5). $\qquad\square$

To simplify the notation we consider the continuous time Markov chain

$$Y = \{Y_t, t \in R^+\} = \{\overline{Y}_{2t}, t \in R^+\}, \tag{2.13}$$

which has potential given by the matrix $V = \{V_{j,k}; j, k \in \overline{\mathbb{N}}\}$ with,

$$V_{j,k} = s_j \wedge s_k = E\left(B_{s_j}B_{s_k}\right), \tag{2.14}$$

where $\{B_t, t \in R^+\}$ is standard Brownian motion, and Q matrix,

$$Q(s) = 2\overline{Q}(s). \tag{2.15}$$

One of our goals is to study permanental processes with kernels of the form (1.12). To that end we now describe the finite potential functions and excessive functions of Y.

Theorem 2.2 *A potential function* $f = Vh$ *is finite if and only if* $h \in \ell_1^+$. *When this is the case the following equivalent conditions hold:*

(i)

$$\frac{f_n - f_{n-1}}{s_n - s_{n-1}} \downarrow 0, \tag{2.16}$$

where we take $f_0 = s_0 = 0$.

(ii) the function $g(s_n) = f_n$ is concave on $\{0\} \cup \{s_j, \ j \geq 1\}$ and

$$\frac{f_n}{s_n} \downarrow 0. \tag{2.17}$$

Proof We point out in the second paragraph following Theorem 1.1 that $f = Vh$ is finite when $h \in \ell_1^+$. The reverse implication follows from the fact that

$$f_1 = \sum_{k=1}^{\infty} V_{1,k} h_k = s_1 \sum_{k=1}^{\infty} h_k, \tag{2.18}$$

where we use (2.14).

In general we have

$$f_n = \sum_{k=1}^{n} s_k h_k + s_n \sum_{k=n+1}^{\infty} h_k, \tag{2.19}$$

and

$$f_{n+1} = \sum_{k=1}^{n} s_k h_k + s_{n+1} \sum_{k=n+1}^{\infty} h_k. \tag{2.20}$$

Therefore,

$$\frac{f_{n+1} - f_n}{s_{n+1} - s_n} = \sum_{k=n+1}^{\infty} h_k. \tag{2.21}$$

This and (2.18) gives *(i)*. It also shows that $g(s_n) = f_n$ is a concave function on $\{0\} \cup \{s_j, \ j \geq 1\}$.

Note that if we divide (2.19) by s_n and (2.20) by s_{n+1}, and use the fact that s_j is strictly increasing, we have

$$\frac{f_{n+1}}{s_{n+1}} < \frac{f_n}{s_n}. \tag{2.22}$$

This shows that if $f = Vh$ then $f_n/s_n \downarrow$.

To see that *(i)* implies *(ii)* set

$$\delta_j = \frac{f_j - f_{j-1}}{s_j - s_{j-1}} \geq 0. \tag{2.23}$$

We write

$$\frac{f_n}{s_n} = \frac{f_p + (f_n - f_p)}{s_p + (s_n - s_p)} = \frac{f_p + \sum_{j=p+1}^{n}(f_j - f_{j-1})}{s_p + (s_n - s_p)} \tag{2.24}$$

$$= \frac{f_p + \sum_{j=p+1}^{n}\delta_j(s_j - s_{j-1})}{s_p + (s_n - s_p)} \leq \frac{f_p + \delta_p \sum_{j=p+1}^{n}(s_j - s_{j-1})}{s_p + (s_n - s_p)}$$

$$= \frac{f_p + \delta_p(s_n - s_p)}{s_p + (s_n - s_p)}.$$

Consequently,

$$\limsup_{n\to\infty} \frac{f_n}{s_n} \leq \delta_p. \tag{2.25}$$

Since this holds for all p we see that (i) implies (ii).

That (ii) implies (i) is an elementary property of concave functions. □

We now describe the finite excessive functions for Y. These are the finite functions f for which $-Q(s)f \geq 0$.

Lemma 2.1 *The function f is a finite excessive function for Y if and only if,*

$$\frac{f_n - f_{n-1}}{s_n - s_{n-1}} \downarrow \delta \geq 0, \tag{2.26}$$

where we take $f_0 = s_0 = 0$.

Proof For all $m \geq 1$,

$$-(Q(s)f)_m = -a_m f_{m-1} + (a_m + a_{m+1}) f_m - a_{m+1} f_{m+1} \tag{2.27}$$
$$= a_m(f_m - f_{m-1}) - a_{m+1}(f_{m+1} - f_m)$$
$$= \frac{f_m - f_{m-1}}{s_m - s_{m-1}} - \frac{f_{m+1} - f_m}{s_{m+1} - s_m}.$$

Since $-Q(s)f \geq 0$, this shows that $(f_m - f_{m-1})/(s_m - s_{m-1})$ is decreasing and consequently has a positive limit which we denote by δ. □

We know that unless $\delta = 0$, f is not a potential function.

We sum up these results in the following lemma:

Lemma 2.2 *Let f is a finite excessive function for Y and set $g(0) = 0$ and*

$$g(s_j) = f_j, \quad j \in \overline{\mathbb{N}}, \tag{2.28}$$

then g is a concave function on $\{0\} \cup \{s_j, \ j \geq 1\}$.

If in addition the function f is a finite potential function for Y then $g(s_j) = o(s_j)$, as $j \to \infty$.

The function $f = \{f_j\}$, where

$$f_j = \delta s_j, \qquad \forall\, j \in \overline{\mathbb{N}}, \tag{2.29}$$

is an excessive function for Y, (in fact $Q(\mathbf{s})f \equiv 0$), but it is not a potential function for Y.

Proof The first statement follows because the terms in (2.27) are positive.

The second statement follows from Theorem 2.2, (ii).

Obviously $Q(\mathbf{s})f \equiv 0$ so f in (2.29) is an excessive function for V. It follows from the second statement that it is not a potential function. $\square$

Lemma 2.3 *Let f be a finite excessive function for Y such that,*

$$\frac{f_n - f_{n-1}}{s_n - s_{n-1}} \downarrow 0, \tag{2.30}$$

where we take $f_0 = s_0 = 0$. Then $f = Vh$ where $h = -Q(\mathbf{s})f \in \ell_1^+$.

Proof Since f is finite and excessive,

$$h_k = (-Q(\mathbf{s})f)_k \geq 0, \qquad \forall\, k \in \overline{\mathbb{N}}. \tag{2.31}$$

By (2.27) and (2.30) we see that

$$\|h\|_1 = \frac{f_1}{s_1}. \tag{2.32}$$

It remains to show that $f = Vh$. Using (2.27), and setting $f_0 = s_0 = 0$, we have that for $n \geq 2$,

$$\sum_{k=1}^{n-1} s_k h_k = \sum_{k=1}^{n-1} s_k \left(\frac{f_k - f_{k-1}}{s_k - s_{k-1}} - \frac{f_{k+1} - f_k}{s_{k+1} - s_k} \right) \tag{2.33}$$

$$= f_1 + \sum_{k=2}^{n-1} (s_k - s_{k-1}) \left(\frac{f_k - f_{k-1}}{s_k - s_{k-1}} \right) - s_{n-1} \left(\frac{f_n - f_{n-1}}{s_n - s_{n-1}} \right)$$

$$= f_{n-1} - s_{n-1} \left(\frac{f_n - f_{n-1}}{s_n - s_{n-1}} \right).$$

Furthermore. using (2.27) and (2.30) we see that,

$$\sum_{k=n}^{\infty} h_k = \frac{f_n - f_{n-1}}{s_n - s_{n-1}}. \tag{2.34}$$

Consequently, that for $n \geq 2$,

$$(Vh)_n = \sum_{k=1}^{n-1} s_k h_k + s_n \sum_{k=n}^{\infty} h_k \tag{2.35}$$

$$= f_{n-1} - s_{n-1}\left(\frac{f_n - f_{n-1}}{s_n - s_{n-1}}\right) + s_n \frac{f_n - f_{n-1}}{s_n - s_{n-1}} = f_n.$$

In addition, by (2.32),

$$(Vh)_1 = s_1 \sum_{k=1}^{\infty} h_k = f_1. \tag{2.36}$$

$\square$

The next corollary sums up the results of Theorem 2.2 and the following lemmas.

Corollary 2.1 *Let $f \geq 0$ be a finite function. Then $f = Vh$ where $h \in \ell_1^+$ if and only if*

$$\frac{f_n - f_{n-1}}{s_n - s_{n-1}} \downarrow 0, \tag{2.37}$$

where we take $f_0 = s_0 = 0$.

Proof It follows from Theorem 2.2 that if $f = Vh$ for some $h \in \ell_1^+$ then f is finite and (2.37) holds. It follows from Lemma 2.1 that (2.37) implies that f is a finite excessive function for Y. Therefore, using Lemma 2.3 we see that $f = Vh$ for some $h \in \ell_1^+$. $\square$

We have the following Riesz decomposition theorem for functions which are excessive for Y.

Theorem 2.3 *Let f be a finite excessive function for Y. Then, necessarily, f satisfies (2.26) for some $\delta \geq 0$, and*

$$f_n = \tilde{f}_n + \delta s_n, \quad \text{for all } n \in \overline{\mathbb{N}}, \tag{2.38}$$

where $\{\tilde{f}_n, n \in \overline{\mathbb{N}}\}$ is a potential function for Y.

Proof Let f be an excessive function for Y and define

$$\tilde{f}_n = f_n - \delta s_n \tag{2.39}$$

for $\delta \geq 0$ as defined in (2.26). This implies that $\widetilde{f}$ is a finite excessive function for Y, and

$$\frac{\widetilde{f}_{n+1} - \widetilde{f}_n}{s_{n+1} - s_n} = \frac{f_{n+1} - f_n}{s_{n+1} - s_n} - \delta, \tag{2.40}$$

which together with (2.26) gives

$$\lim_{n \to \infty} \frac{\widetilde{f}_n - \widetilde{f}_{n-1}}{s_n - s_{n-1}} \downarrow 0. \tag{2.41}$$

By Lemma 2.3 we see that $\widetilde{f}$ is a potential function for Y.   □

We now consider the asymptotic properties of permanental processes with kernels that are not symmetric but are modifications of symmetric potentials. Let $\widetilde{Y}_\alpha = \{\widetilde{Y}_{\alpha,n}, n \in \overline{\mathbb{N}}\}$ be a permanental processes with kernel $\widetilde{V} = \{\widetilde{V}_{j,k}, j, k \in \overline{\mathbb{N}}\}$ where,

$$\widetilde{V}_{j,k} = s_j \wedge s_k + f_k. \tag{2.42}$$

and $f = \{f_k, k \in \overline{\mathbb{N}}\}$ is a finite potential function for Y.

Since we use Theorem 1.2 to find the asymptotic behavior of $\widetilde{Y}_\alpha$ we need only deal with finite sections of kernels.

Lemma 2.4 Let $V(1, n)$ be an $n \times n$ matrix with elements

$$V(1, n)_{j,k} = s_j \wedge s_k \qquad j, k = 1, \ldots, n, \tag{2.43}$$

in which s_j is a strictly increasing sequence. Then

$$V(1, n)^{-1} = \begin{pmatrix} a_1 + a_2 & -a_2 & 0 & \cdots & 0 & 0 \\ -a_2 & a_2 + a_3 & -a_3 & \cdots & 0 & 0 \\ \vdots & \vdots & \vdots & \ddots & \vdots & \vdots \\ 0 & 0 & 0 & \cdots & a_{n-1} + a_n & -a_n \\ 0 & 0 & 0 & \cdots & -a_n & a_n \end{pmatrix}, \tag{2.44}$$

where $\{a_j\}$ is given in (2.2).

Proof It is easy to verify that this is the inverse of $V(1, n)$.   □

Note that the first n-1 rows of $V(1, n)^{-1}$ are the same as the first n-1 rows of the matrix in (1.30).

Lemma 2.5 Let $V(l, n)$ be an $n \times n$ matrix with elements

$$V(l, n)_{j,k} = s_j \wedge s_k \qquad j, k = l + 1, \ldots, l + n, \tag{2.45}$$

in which s_j is a strictly increasing sequence. Then
$V(l, n)^{-1} =$

$$
\begin{pmatrix}
1/s_{l+1} + a_{l+2} & -a_{l+2} & 0 & \cdots & 0 & 0 \\
-a_{l+2} & a_{l+2} + a_{l+3} & -a_{l+3} & \cdots & 0 & 0 \\
\vdots & \vdots & \vdots & \ddots & \vdots & \vdots \\
0 & 0 & 0 & \cdots & a_{l+n-1} + a_{l+n} & -a_{l+n} \\
0 & 0 & 0 & \cdots & -a_{l+n} & a_{l+n}
\end{pmatrix}.
\tag{2.46}
$$

Proof This follows immediately from Lemma 2.4 by relabeling the a. and taking $a_1 = 1/s_{l+1}$. An alternate proof is simply to verify that (2.46) is the inverse of $V(l, n)$. □

In the next lemma we give the estimate that enables us to apply Theorem 1.2. Recall that for any invertible matrix M we often denote $M^{-1}_{j,k}$ by $M^{j,k}$.

Lemma 2.6 *Let f be a potential function for Y. Then*

$$
\sum_{j,k=1}^{n} V(l, n)^{k,j} f_{l+j} = o_l(1),
\tag{2.47}
$$

uniformly in n.

Proof Note that

$$
\sum_{j,k=1}^{n} V(l, n)^{k,j} f_{l+j} = \sum_{j=1}^{n} f_{l+j} \sum_{k=1}^{n} V(l, n)^{k,j} = \frac{f_{l+1}}{s_{l+1}},
\tag{2.48}
$$

where we use the fact that all the column sums of $V(l, n)^{-1}$ are equal to zero except for the first one. Therefore, (2.47) follows from (2.17). □

Proof of Theorem 1.6 We first use Theorem 1.2. Therefore, we need to obtain the denominator in (1.17) for the Gaussian sequence $\xi = \{\xi_j, j \in \overline{\mathbb{N}}\}$ where,

$$
E(\xi_j \xi_k) = s_j \wedge s_k, \qquad j, k \in \overline{\mathbb{N}}.
\tag{2.49}
$$

We use Koval's Theorem, [9, page 1]. This involves the function

$$
\mathcal{K}_{s_i, M}(j) = \log \left(\sum_{i=1}^{j-1} M \wedge \log (s_{i+1}/s_i) \right),
\tag{2.50}
$$

for any number $M > 0$. (Note that in the notation introduced in (1.35), $\mathcal{K}_{s_i}(j) = \mathcal{K}_{s_i,1}(j)$.) Since for any $M > 0$,

$$\lim_{j \to \infty} \frac{\mathcal{K}_{s_i}(j)}{\mathcal{K}_{s_i,M}(j)} = 1, \tag{2.51}$$

we use $\mathcal{K}_{s_i}(j)$ to avoid ambiguity.

Koval's Theorem states that

$$\overline{\lim_{j \to \infty}} \frac{\xi_j}{(2s_j \mathcal{K}_{s_i}(j))^{1/2}} = 1, \qquad a.s. \tag{2.52}$$

Note that for any $M > 0$,

$$\lim_{j \to \infty} \mathcal{K}_{s_i,M}(j) = \infty. \tag{2.53}$$

This is obvious when $\limsup_{i \to \infty} \log(s_{i+1}/s_i) > M$ because there would be an infinite number of the terms M in the sum. If $\limsup_{i \to \infty} \log(s_{i+1}/s_i) \leq M$, then replacing M by $2M$, we can find an i_0 such that,

$$\sum_{i=1}^{j-1} 2M \wedge \log(s_{i+1}/s_i) > \sum_{i=i_0}^{j-1} \log(s_{i+1}/s_i) = \log s_j - \log s_{i_0}. \tag{2.54}$$

By (2.51), $\lim_{j \to \infty} \mathcal{K}_{s_i,2M}(j)/\mathcal{K}_{s_i,M}(j) = 1$ so we get (2.53).

By Theorem 1.1, V is the kernel of α-permanental processes for all $\alpha > 0$. In addition we see by Lemma 2.6 that (1.16) is satisfied, and by (2.17) and (2.53) that (1.18) is satisfied. Consequently, we can use Theorem 1.2 to get (1.37) for all $1/2 \leq \alpha < \infty$. Since $\tilde{Y}_\alpha$ is infinitely divisible and positive, it is obvious that the upper bound in (1.37) holds for all $\alpha > 0$.

We now show that the lower bounds in (1.37) holds for all $\alpha > 0$. To show this it suffices to find a subsequence $\{s_{p_j}\}$ of $\{s_j\}$ such that

$$\limsup_{j \to \infty} \frac{\tilde{Y}_{\alpha, p_j}}{s_{p_j} \mathcal{K}_{s_i}(p_j)} \geq 1, \qquad a.s., \qquad \forall \alpha > 0. \tag{2.55}$$

We choose $\{s_{p_j}\}$ recursively as follows:

$$s_{p_{j+1}} = \min \left\{ k : \frac{s_k}{s_{p_j}} \geq \theta \right\} \tag{2.56}$$

where $s_{p_1} = 1$ and $\theta \geq e$. Clearly

$$\frac{s_{p_{j+1}}}{s_{p_j}} \geq \theta, \qquad \text{and} \qquad \frac{s_{p_{j+1}-1}}{s_{p_j}} < \theta. \tag{2.57}$$

Consequently,

$$\sum_{k=p_l}^{p_{l+1}-1} 1 \wedge \log \frac{s_{k+1}}{s_k} \leq \left(\sum_{k=p_l}^{p_{l+1}-2} \log \frac{s_{k+1}}{s_k} \right) + 1 \tag{2.58}$$

$$\leq \log \theta + 1 < 2 \log \theta.$$

It follows from these relationships that,

$$\mathcal{K}_{s_i}(p_j) = \log \left(\sum_{i=1}^{p_j-1} 1 \wedge \log \frac{s_{i+1}}{s_i} \right) \tag{2.59}$$

$$= \log \left(\sum_{l=1}^{j} \sum_{i=p_l}^{p_{l+1}-1} 1 \wedge \log \frac{s_{i+1}}{s_i} \right)$$

$$\leq \log j + \log \log \theta^2.$$

This shows that

$$\limsup_{j \to \infty} \frac{\widetilde{Y}_{\alpha,p_j}}{s_{p_j} \mathcal{K}_{s_i}(p_j)} \geq \limsup_{j \to \infty} \frac{\widetilde{Y}_{\alpha,p_j}}{s_{p_j} \log j}. \tag{2.60}$$

Therefore, to obtain (2.55) it suffices to show that

$$\limsup_{j \to \infty} \frac{\widetilde{Y}_{\alpha,p_j}}{s_{p_j} \log j} \geq 1. \tag{2.61}$$

To do this we first extend and relabel $\widetilde{Y}_\alpha^{(p)} = \{\widetilde{Y}_{\alpha,p_j}, j \in \overline{\mathbb{N}}\}$ to the permanental process $\widehat{Y}_\alpha^{(p)} = \{\widehat{Y}_{\alpha,j}, j \in \{0\} \cup \overline{\mathbb{N}}\}$ with kernel,

$$K_{j,k} = s_{p_j} \wedge s_{p_k} + f(s_{p_k}), \qquad j, k \in \overline{\mathbb{N}}, \tag{2.62}$$

$$K_{0,0} = 1, \qquad K_{j,0} = 1, \qquad j \in \overline{\mathbb{N}}, \qquad \text{and} \qquad K_{0,k} = f(s_{p_k}), \qquad k \in \overline{\mathbb{N}}.$$

It is clear that $\widehat{Y}_\alpha^{(p)} \overset{law}{=} \widetilde{Y}_\alpha^{(p)}$ on $\overline{\mathbb{N}}$, so that to obtain (2.61) it suffices to show that,

$$\limsup_{j \to \infty} \frac{\widehat{Y}_{\alpha,j}}{s_{p_j} \log j} \geq 1. \tag{2.63}$$

(Note that by definition, to show that $\widehat{Y}_\alpha^{(p)}$ is a permanental process it suffices to show that for all $\{i_1, \ldots, i_n\} \in \{0\} \cup \overline{\mathbb{N}}$, $\{K_{i_j,i_k}\}_{j,k=0}^n$ is the kernel of a permanental process. It follows as in in (6.8)–(6.11) that $\{K_{i_j,i_k}\}_{j,k=0}^n$ is an inverse M-matrix. Hence by [5, Lemma 4.2] it is the kernel of a permanental process.)

Let $V^{(p)} = \{V^{(p)}_{j,k}, j, k \in \overline{\mathbb{N}}\}$ where,

$$V^{(p)}_{j,k} = s_{p_j} \wedge s_{p_k}. \tag{2.64}$$

Let $K(0, n+1)$ denote the matrix $\{K_{j,k}\}^n_{j,k=0}$. It follows from (6.11) that for $j \geq 1$ the reciprocal of the diagonal element of the j-th row of $(K(0, n+1))^{-1}$, i.e., $1/K(0, n+1)^{j,j}$, satisfies,

$$1/K(0, n+1)^{j,j} = 1/V^{(p)}(1, n)^{j,j}, \qquad 1 \leq j \leq n. \tag{2.65}$$

It follows from Lemma 2.4 with s_j replaced by s_{p_j}, and the second equality in (2.11), that for $1 \leq j < n$,

$$1/V^{(p)}(1, n)^{j,j} = \frac{\left(s_{p_{j+1}} - s_{p_j}\right)\left(s_{p_j} - s_{p_{j-1}}\right)}{s_{p_{j+1}} - s_{p_{j-1}}} \tag{2.66}$$

$$= s_{p_j} \left(\frac{\left(1 - s_{p_j}/s_{p_{j+1}}\right)\left(1 - s_{p_{j-1}}/s_{p_j}\right)}{1 - s_{p_{j-1}}/s_{p_{j+1}}}\right)$$

$$\geq s_{p_j} \left(1 - s_{p_j}/s_{p_{j+1}}\right)\left(1 - s_{p_{j-1}}/s_{p_j}\right).$$

Using (2.57) we see that for $1 \leq j < n$,

$$1/K(0, n+1)^{j,j} \geq s_{p_j}(1 - 1/\theta)^2. \tag{2.67}$$

Since this holds for all n, it follows from [15, Lemma 7.3] that,

$$\varlimsup_{j \to \infty} \frac{\widehat{Y}_{\alpha,j}}{s_{p_j} \log j} \geq (1 - (1/\theta))^2 \qquad a.s., \tag{2.68}$$

and since we can take θ arbitrarily large we get (2.63).   □

We continue to study the behavior of the function $\mathcal{K}_{s_i}(j)$.

Lemma 2.7

$$\mathcal{K}_{s_i}(j) \leq \log \log s_j \wedge \log j. \tag{2.69}$$

Furthermore, if

$$\liminf_{i \to \infty} \frac{s_i}{s_{i-1}} > 1, \qquad then \qquad \lim_{j \to \infty} \frac{\mathcal{K}_{s_i}(j)}{\log j} = 1 \tag{2.70}$$

and if

$$\limsup_{i \to \infty} \frac{s_i}{s_{i-1}} < \infty \qquad then \qquad \lim_{j \to \infty} \frac{\mathcal{K}_{s_i}(j)}{\log \log s_j} = 1. \tag{2.71}$$

Proof The statement in (2.69) is trivial. To continue, consider $\mathcal{K}_{s_i,M}(j)$ in (2.50). If $\liminf_{i\to\infty}\frac{s_i}{s_{i-1}} > 1$ holds there exist numbers $m_0 > 0$ and i_0 such that,

$$\inf_{i\geq i_0} \frac{s_i}{s_{i-1}} \geq e^{m_0}, \tag{2.72}$$

which implies that,

$$\inf_{i\geq i_0} \log(s_i/s_{i-1}) > m_0 > 0. \tag{2.73}$$

Therefore,

$$\lim_{j\to\infty} \frac{\mathcal{K}_{s_i,m_0}(j)}{\log j} = 1, \tag{2.74}$$

which, by (2.51) gives (2.70).

To get (2.71) we simply take $M = 1 + \log D$ in (2.50), where $D = \limsup_{i\to\infty} s_i/s_{i-1}$. $\quad\square$

When $\liminf_{i\to\infty} s_i/s_{i-1} = 1$, we can't simplify $\mathcal{K}_{s_i}(j)$ without imposing additional conditions. It can oscillate between $\log j$ and $\log\log s_j$ when $\log\log s_j < \log j$. (Of course it is possible that $\log\log s_j > \log j$ for some j, or even for most j, but because of (2.69) we needn't be concerned with these cases.)

We can be more precise when,

$$\lim_{j\to\infty} \frac{s_{j+1}}{s_j} = 1. \tag{2.75}$$

We can write,

$$s_j = \exp\left(\sum_{k=1}^{j} \epsilon_k\right), \quad \text{where } \epsilon_k > 0, \lim_{k\to\infty} \epsilon_k = 0, \tag{2.76}$$

and the sum diverges. Since $\epsilon_k \to 0$ we have $\log\log s_j < \log j$ for all j sufficiently large, but we may still have

$$\lim_{j\to\infty} \frac{\log\log s_j}{\log j} = 1. \tag{2.77}$$

This is the case if $\epsilon_k = 1/\log k$, which implies that $s_j \sim \exp(j/\log j)$, as $j \to \infty$, and the right-hand side of (2.70) still holds.

We give some more examples.

Example 2.1 (i) If $\epsilon_k = k^{\alpha-1}$, for $0 < \alpha < 1$, we have $s_j \sim \exp(j^\alpha/\alpha)$, as $j \to \infty$, and,

$$\lim_{j\to\infty} \frac{\log\log s_j}{\log j} = \alpha. \tag{2.78}$$

Consequently, by (2.71),

$$\lim_{j \to \infty} \frac{\mathcal{K}_{s_i}(j)}{\log \log s_j} = \lim_{j \to \infty} \frac{\mathcal{K}_{s_i}(j)}{\alpha \log j} = 1. \tag{2.79}$$

(ii) If $\epsilon_k = k^{-1}$, we have $s_j \sim j$ as $j \to \infty$, and,

$$\lim_{j \to \infty} \frac{\mathcal{K}_{s_i}(j)}{\log \log s_j} = \lim_{j \to \infty} \frac{\mathcal{K}_{s_i}(j)}{\log \log j} = 1. \tag{2.80}$$

(iii) If $\epsilon_k = 1/(k \log k)$, we have $s_j \sim \log j$ as $j \to \infty$, and,

$$\lim_{j \to \infty} \frac{\mathcal{K}_{s_i}(j)}{\log \log s_j} = \lim_{j \to \infty} \frac{\mathcal{K}_{s_i}(j)}{\log \log \log j} = 1. \tag{2.81}$$

Chapter 3
Birth and Death Processes with Emigration

A continuous time birth and death process with emigration is a Markov chain with a tridiagonal Q matrix. When all the row sums of the Q matrix, except for the first row sum, are equal to zero, it is called, simply, a birth and death process. In this section we generalize the Q matrix $Q(\mathbf{s})$ defined in (2.15) to get a large class of Q matrices of continuous time birth and death process with emigration.

For any sequence $b = (b_1, b_2, \ldots)$ define $D_b = \mathrm{diag}\,(b_1, b_2, \ldots)$. We have the following obvious but important lemma:

Lemma 3.1 *Let Q denote the Q-matrix of a Markov chain Y on $\overline{\mathbb{N}}$. If b is an excessive function for Y, then*

$$D_b Q D_b \tag{3.1}$$

is also a Q-matrix.

Proof This follows immediately since b is positive and $Qb \leq 0$. $\qquad\square$

We apply Lemma 3.1 to $Q(\mathbf{s})$ defined in (2.15). We point out in the paragraph containing (2.15) that $Q(\mathbf{s})$ is the Q-matrix of a continuous time Markov chain Y with potential $V = \{V_{j,k}, j, k \in \overline{\mathbb{N}}\}$ where,

$$V_{j,k} = s_j \wedge s_k, \qquad j, k \in \overline{\mathbb{N}}, \tag{3.2}$$

and $\mathbf{s} = \{s_j, j \geq 1\}$ is a strictly increasing sequence with $s_j > 0$ and $\lim_{j \to \infty} s_j = \infty$. The next lemma is a significant generalization of this observation.

Lemma 3.2 *Let $Z = \{Z_t, t \in R^+\}$ be a continuous symmetric transient Markov chain on $\overline{\mathbb{N}}$ with Q matrix $D_b Q(\mathbf{s}) D_b$, where b is a finite potential function for the Markov chain Y defined in (2.13). Then $W = \{W_{j,k}, j, k \in \overline{\mathbb{N}}\}$ where*

© The Author(s), under exclusive license to Springer Nature Switzerland AG 2021
M. B. Marcus and J. Rosen, *Asymptotic Properties of Permanental Sequences*,
SpringerBriefs in Probability and Mathematical Statistics,
https://doi.org/10.1007/978-3-030-69485-2_3

$$W_{j,k} = \frac{1}{b_j} V_{j,k} \frac{1}{b_k}, \tag{3.3}$$

is the potential for Z.

Remark 3.1 In Lemma 3.2 we take b to be a finite potential function for Y. It follows from Theorem 2.2 that the function $g(s_j) = b_j$ is an increasing concave function of $\{s_j\}$ and $s_j/b_j \uparrow \infty$.

Consider $\{f_j\}$, the finite potential functions of W. We have

$$f_j = \sum_{k=1}^{\infty} W_{j,k} h_k = \sum_{k=1}^{\infty} \frac{1}{b_j} V_{j,k} \frac{h_k}{b_k}. \tag{3.4}$$

Consequently

$$b_j f_j = b_j \sum_{k=1}^{\infty} W_{j,k} h_k = \sum_{k=1}^{\infty} V_{j,k} \frac{h_k}{b_k}. \tag{3.5}$$

Therefore, $\{b_j f_j\}$, is a finite potential function for Y. As noted in the first paragraph of this remark this implies that $g(s_j) = b_j$ is an increasing concave function of $\{s_j\}$. Therefore we can write f as

$$f_j = \frac{g(s_j)}{h(s_j)}, \qquad \forall j \in \overline{\mathbb{N}}, \tag{3.6}$$

where g and h are positive strictly concave functions.

Proof of Lemma 3.2 It is easy to see that $Q(s)V = -I$ in the sense of multiplication of infinite matrices. Consequently, since $D_b W = V D_b^{-1}$, it follows that we also have

$$D_b Q(s) D_b W = D_b Q(s) V D_b^{-1} = -I. \tag{3.7}$$

Let $\overline{W}$ be the potential for Z. Using Lemma 8.1 we see that $D_b Q(s) D_b \overline{W} = -I$. Consequently,

$$Q(s) D_b \left(W - \overline{W} \right) = 0. \tag{3.8}$$

Consider the equation $Q(s)g = 0$. Using (2.27) we see that we must have

$$\frac{g_j - g_{j-1}}{s_j - s_{j-1}} = c_0 \qquad \forall j \geq 1, \tag{3.9}$$

for some fixed constant c_0 where we set $g_0 = s_0 = 0$. Therefore, all solutions of $Q(s)g = 0$ are of the form $g = c_0(s_1, s_2, \ldots)$.

Consider the components of (3.8). We see that for all $k \in \overline{\mathbb{N}}$,

$$\sum_j (Q(s))_{l,j} (D_b(W - \overline{W}))_{j,k} = 0. \tag{3.10}$$

Therefore, using the observations in the preceding paragraph, we have that for each $k \in \overline{\mathbb{N}}$,

$$W_{j,k} - \overline{W}_{j,k} = c_k s_j / b_j, \qquad \forall j \geq 1, \tag{3.11}$$

for some constant c_k.

We now show that $c_k = 0$ for all k. Let P^j denote probabilities for Z. We have

$$\overline{W}_{j,k} = P^j (T_k < \infty) \overline{W}_{k,k}. \tag{3.12}$$

Using this and (3.11) we see that,

$$\frac{s_k}{b_j b_k} = W_{j,k} = P^j (T_k < \infty) \overline{W}_{k,k} + c_k s_j / b_j, \qquad \forall j \geq k. \tag{3.13}$$

Since b_j is increasing and $s_j b_j \uparrow \infty$, this is only possible if $c_k = 0$. $\qquad\square$

Remark 3.2 This Lemma also applies if b is a general finite excessive function for Y. That is, by Theorem 2.3, if we add δs_j to the present b_j. In that case, the left-hand side of (3.13) goes to zero as $j \to \infty$, and the last term in (3.13) converges to c_k / δ, which again shows that $c_k = 0$.

Remark 3.3 We also note that if $Y = \{Y_t, t \in R^+\}$ is a process with Q-matrix $Q(s)$, then Z can be obtained from Y by first doing a b-transform and then a time change by the inverse of the continuous additive functional $\int_0^t b(Y_s)^{-2} ds$. This gives an alternate proof of Lemma 3.2.

Our goal in this section is to use Theorem 1.2 to prove Theorem 1.7. We use the next two lemmas to obtain (1.16).

Lemma 3.3 *Let* $f = Wh$, *where* $h \in \ell_1^+$. *Then*

$$\sum_{j,k=1}^{n} (W(l,n))^{j,k} f_{k+l} = \frac{f_{l+1}}{W_{l+1,l+1}} + o_l(1), \quad \text{uniformly in } n. \tag{3.14}$$

Proof For any $l, k \in \overline{\mathbb{N}}$ we set $f_k^{(l)} = f_{l+k}$. Similarly, we set $a_k^{(l)} = a_{l+k}$ and $b_k^{(l)} = b_{l+k}$. We have

$$\sum_{j,k=1}^{n} W(l,n)^{j,k} f_k^{(l)} = \sum_{k=1}^{n} f_k^{(l)} b_k^{(l)} \sum_{j=1}^{n} (V(l,n))^{k,j} b_j^{(l)}. \tag{3.15}$$

For any sequence $\{c_k\}$ we use the standard notation $\Delta c_k = c_{k+1} - c_k$.

Using (2.46) we see that,

$$\sum_{j=1}^{n} (V(l,n))^{1,j} b_j^{(l)} = \frac{b_1^{(l)}}{s_1^{(l)}} - \Delta b_1^{(l)} a_2^{(l)}, \tag{3.16}$$

and for $1 < k < n$,

$$\sum_{j=1}^{n}(V(l,n))^{k,j}b_j^{(l)} = -b_{k-1}^{(l)}a_k^{(l)} + b_k^{(l)}(a_k^{(l)} + a_{k+1}^{(l)}) - b_{k+1}^{(l)}a_{k+1}^{(l)} \tag{3.17}$$

$$= \Delta b_{k-1}^{(l)}a_k^{(l)} - \Delta b_k^{(l)}a_{k+1}^{(l)},$$

and

$$\sum_{j=1}^{n}(V(l,n))^{n,j}b_j^{(l)} = \Delta b_{n-1}^{(l)}a_n^{(l)}. \tag{3.18}$$

It follows from (3.15)–(3.18) that,

$$\sum_{j,k=1}^{n} W(l,n)^{j,k} f_k^{(l)} = \left(f_1^{(l)}b_1^{(l)}\right)\frac{b_1^{(l)}}{s_1^{(l)}} + \sum_{k=2}^{n} a_k^{(l)}\Delta b_{k-1}^{(l)}\Delta\left(f_{k-1}^{(l)}b_{k-1}^{(l)}\right). \tag{3.19}$$

Set $\widetilde{f} = D_b f$. Then, since $f = Wh$, for some $h \in \ell_1^+$, we see that

$$\widetilde{f}_k = b_k \sum_{j=1}^{\infty} W_{k,j}h_j = \sum_{j=1}^{\infty} V_{k,j}\frac{h_j}{b_j}, \qquad \forall k \in \overline{\mathbb{N}}. \tag{3.20}$$

This shows that $\widetilde{f} = V(D_b^{-1}h)$. Therefore, by (2.21), we see that for all $n \geq 1$,

$$\frac{\widetilde{f}_{n+1} - \widetilde{f}_n}{s_{n+1} - s_n} = \sum_{k=n+1}^{\infty} \frac{h_k}{b_k}. \tag{3.21}$$

We now use (3.19) and (3.21) and the fact that $\widetilde{f} = D_b f$ to get,

$$\sum_{j,k=1}^{n} W(l,n)^{j,k} f_k^{(l)} = f_1^{(l)}\frac{\left(b_1^{(l)}\right)^2}{s_1^{(l)}} + \sum_{j=2}^{n} a_j^{(l)}\Delta b_{j-1}^{(l)}\Delta\widetilde{f}_{j-1}^{(l)} \tag{3.22}$$

$$= f_1^{(l)}\frac{\left(b_1^{(l)}\right)^2}{s_1^{(l)}} + \sum_{j=2}^{n} \frac{\Delta b_{j-1}^{(l)}\Delta\widetilde{f}_{j-1}^{(l)}}{s_j^{(l)} - s_{j-1}^{(l)}}$$

$$= \frac{f_{l+1}}{W_{l+1,l+1}} + \sum_{j=2}^{n} \Delta b_{j-1}^{(l)}\sum_{k=j}^{\infty}\frac{h_k^{(l)}}{b_k^{(l)}}.$$

Since

$$\sum_{j=2}^{n} \Delta b_{j-1}^{(l)} \sum_{k=j}^{\infty} \frac{h_k^{(l)}}{b_k^{(l)}} \tag{3.23}$$

$$= -b_1^{(l)} \sum_{k=2}^{\infty} \frac{h_k^{(l)}}{b_k^{(l)}} + \sum_{j=2}^{n-1} b_j^{(l)} \left(\sum_{k=j}^{\infty} \frac{h_k^{(l)}}{b_k^{(l)}} - \sum_{k=j+1}^{\infty} \frac{h_k^{(l)}}{b_k^{(l)}} \right) + b_n^{(l)} \sum_{k=n}^{\infty} \frac{h_k^{(l)}}{b_k^{(l)}}$$

$$\leq \sum_{j=1}^{n-1} h_j^{(l)} + b_n^{(l)} \sum_{k=n}^{\infty} \frac{h_k^{(l)}}{b_k^{(l)}} \leq \sum_{j=1}^{n-1} h_j^{(l)} + \sum_{k=n}^{\infty} h_k^{(l)} = \sum_{j=1}^{\infty} h_j^{(l)},$$

we get (3.14). □

Using the next lemma with Lemma 3.3 we get (1.16).

Lemma 3.4 Let $f = Wh$, $h \in \ell_1^+$. Then

$$\lim_{j \to \infty} \frac{f_j}{W_{j,j}} = 0. \tag{3.24}$$

Proof For $p < j + 2$, we have,

$$f_j = \sum_{k=1}^{\infty} W_{j,k} h_k = \frac{1}{b_j} \sum_{k=1}^{\infty} V_{j,k} \frac{h_k}{b_k} \tag{3.25}$$

$$= \frac{1}{b_j} \sum_{k=1}^{p} \frac{s_k h_k}{b_k} + \frac{1}{b_j} \sum_{k=p+1}^{j-1} \frac{s_k h_k}{b_k} + \frac{s_j}{b_j} \sum_{k=j}^{\infty} \frac{h_k}{b_k}$$

$$\leq \frac{1}{b_j} \frac{s_p}{b_p} \sum_{k=1}^{p} h_k + \frac{s_j}{b_j^2} \sum_{k=p+1}^{\infty} h_k = \frac{1}{b_j} \frac{s_p}{b_p} \sum_{k=1}^{p} h_k + W_{j,j} \sum_{k=p+1}^{\infty} h_k,$$

where for the last line we note that by Remark 3.1, b_k is increasing and $s_k/b_k \uparrow$. Therefore,

$$\frac{f_j}{W_{j,j}} \leq \frac{s_p}{b_p} \frac{b_j}{s_j} \|h\|_1 + \sum_{k=p+1}^{\infty} h_k. \tag{3.26}$$

Using Remark 3.1 again we see that for all $p > 0$,

$$\lim_{j \to \infty} \frac{f_j}{W_{j,j}} \leq \sum_{k=p+1}^{\infty} h_k. \tag{3.27}$$

This gives (3.24). □

Proof of Theorem 1.7 Let $\xi = \{\xi_j, j \in \overline{\mathbb{N}}\}$ be a Gaussian sequence with covariance W. It follows from Koval's Theorem that

$$\limsup_{j \to \infty} \frac{\xi_j}{(2W_{j,j}\mathcal{K}_{s_i}(j))^{1/2}} = 1 \quad a.s. \tag{3.28}$$

Therefore, for $\alpha \geq 1/2$, (1.43) follows from Theorem 1.2. Note that Lemmas 3.3 and 3.4 give (1.16). In addition Lemma 3.4 and (2.53) shows that (1.18) holds. Also, as we have pointed out, the upper bound in (1.43) actually holds for all $\alpha > 0$.

We now show that the lower bounds in (1.43) holds for all $\alpha > 0$. To this it suffices to find a subsequence $\{s_{p_j}\}$ of $\{s_j\}$ such that

$$\limsup_{j \to \infty} \frac{\widetilde{Z}_{\alpha, p_j}}{W_{p_j, p_j} \mathcal{K}_{s_i}(p_j)} \geq 1, \quad a.s., \quad \forall \alpha > 0. \tag{3.29}$$

If we choose $\{s_{p_j}\}$ as in (2.56), this follows if we show that,

$$\limsup_{j \to \infty} \frac{\widetilde{Z}_{\alpha, p_j}}{W_{p_j, p_j} \log j} \geq 1, \quad a.s., \quad \forall \alpha > 0. \tag{3.30}$$

Consider the permanental process $\widetilde{Z}_\alpha^{(p)} = \{\widetilde{Z}_{\alpha, p_j}, j \in \overline{\mathbb{N}}\}$. As in the proof of Theorem 1.6 we extend and relabel $\widetilde{Z}_\alpha^{(p)}$ to get a permanental process $\widehat{Z}_\alpha^{(p)} = \{\widehat{Z}_{\alpha, j}, j \in \{0\} \cup \overline{\mathbb{N}}\}$ with kernel $\overline{K} = \{\overline{K}_{j,k}; j, k \in \{0\} \cup \overline{\mathbb{N}}\}$ where,

$$\overline{K}_{j,k} = \frac{s_{p_j} \wedge s_{p_k}}{b_{p_j} b_{p_k}} + f(s_{p_k}), \quad j, k \in \overline{\mathbb{N}}, \tag{3.31}$$

$$\overline{K}_{0,0} = 1, \quad \overline{K}_{j,0} = 1, \quad j \in \overline{\mathbb{N}}, \quad \text{and} \quad \overline{K}_{0,k} = f(s_{p_k}), \quad k \in \overline{\mathbb{N}}.$$

It is clear that $\widehat{Z}_\alpha^{(p)} \overset{law}{=} \widetilde{Z}_\alpha^{(p)}$ on $\overline{\mathbb{N}}$, so that to obtain (3.30) it suffices to show that,

$$\limsup_{j \to \infty} \frac{\widehat{Z}_{\alpha, j}}{W_{p_j, p_j} \log j} \geq 1. \tag{3.32}$$

Let $\overline{W}^{(p)} = \{\overline{W}_{j,k}^{(p)}, j, k \in \overline{\mathbb{N}}\}$ where,

$$\overline{W}_{j,k}^{(p)} = \frac{s_{p_j} \wedge s_{p_k}}{b_{p_j} b_{p_k}}, \tag{3.33}$$

and let $\overline{K}(0, n+1)$ denote the matrix $\{\overline{K}_{j,k}\}_{j,k=0}^n$. As in the proof of Theorem 1.6, it follows from (6.11) that for $j \geq 1$,

$$1/\overline{K}(0, n+1)^{j,j} = 1/\overline{W}^{(p)}(1, n)^{j,j}, \qquad 1 \le j \le n. \tag{3.34}$$

It is easy to see that

$$\overline{W}^{(p)}(1, n)^{j,j} = b_j^2 V^{(p)}(1, n)^{j,j}, \tag{3.35}$$

where $V^{(p)}$ is given in (2.64). Therefore, analogous to (2.66) and (2.67) we see that,

$$1/\overline{K}(0, n+1)^{j,j} \ge \frac{s_{p_j}}{b_j^2}(1 - 1/\theta)^2 = W_{p_j,p_j}(1 - 1/\theta)^2. \tag{3.36}$$

As in the proof of Theorem 1.6 this implies (3.32). □

Remark 3.4 Let $\widetilde{Y}_\alpha$ be as in Theorem 1.6. The kernel of $\widetilde{Y}_\alpha$ is $\widetilde{V} = \{\widetilde{V}_{j,k}; j, k \in \overline{\mathbb{N}}\}$, where

$$\widetilde{V}_{j,k} = V_{j,k} + f_k, \qquad j, k \ge 1. \tag{3.37}$$

It follows from (1.11) that $Z'_\alpha := D_b^{-2}\widetilde{Y}_\alpha = (b_1^{-2}Y_{\alpha,1}, b_2^{-2}Y_{\alpha,2}, \ldots)$ has kernel $W' = \{W'_{j,k}; j, k \in \overline{\mathbb{N}}\}$ where

$$W'_{j,k} = \frac{V_{j,k}}{b_j b_k} + \frac{f_k}{b_j b_k} = W_{j,k} + \frac{f_k}{b_j b_k}, \qquad j, k \in \overline{\mathbb{N}}. \tag{3.38}$$

This is because for all $n \times n$ matrices K and $D_{b'}^{-1}$,

$$|I + KD_{b'}^{-2}S| = |I + D_{b'}^{-1}KD_{b'}^{-1}S|. \tag{3.39}$$

It follows from Theorem 1.6 that

$$\limsup_{j \to \infty} \frac{b_j^2 Z'_\alpha(j)}{V_{j,j}\mathcal{K}_{s_i}(j)} = 1 \quad a.s., \tag{3.40}$$

or, equivalently

$$\limsup_{j \to \infty} \frac{Z'_\alpha(j)}{W_{j,j}\mathcal{K}_{s_i}(j)} = 1 \quad a.s. \tag{3.41}$$

as in (1.43).

This is easy, but Z'_α has kernel W' whereas $\widetilde{Z}_a$ in Theorem 1.7 has kernel $\widetilde{W}$, in (1.42). In Theorem 1.1 we set out to consider symmetric kernels perturbed by an excessive function f as in (1.12). This is what we do in Theorem 1.7.

In Lemma 3.4 we use the explicit representation of W. It is interesting to note that (3.24) holds in great generality when the diagonals of the matrix go to infinity.

Lemma 3.5 Let $f = Wh$, $h \in \ell_1^+$, for some infinite matrix W such that

$$W_{k,j} \le W_{k,k}, \qquad \forall k \in \overline{\mathbb{N}}. \tag{3.42}$$

Then

$$f_k \le W_{k,k} \, \|h\|_1, \text{ and if } \lim_{k \to \infty} W_{k,k} = \infty, \quad f_k = o(W_{k,k}). \tag{3.43}$$

Proof Let $\epsilon > 0$. We have

$$f_k = \sum_{\{j : W_{j,j} \le \epsilon W_{k,k}\}} W_{k,j} h_j + \sum_{\{j : W_{j,j} > \epsilon W_{k,k}\}} W_{k,j} h_j \tag{3.44}$$

$$\le \epsilon W_{k,k} \sum_{\{j : W_{j,j} \le \epsilon W_{k,k}\}} h_j + W_{k,k} \sum_{\{j : W_{j,j} > \epsilon W_{k,k}\}} h_j$$

$$\le \epsilon W_{k,k} \, \|h\|_1 + W_{k,k} \sum_{\{j : W_{j,j} > \epsilon W_{k,k}\}} h_j.$$

Therefore,

$$\lim_{k \to \infty} \frac{f_k}{W_{k,k}} \le \epsilon \, \|h\|_1 + \lim_{k \to \infty} \sum_{\{j : W_{j,j} > \epsilon W_{k,k}\}} h_j. \tag{3.45}$$

If $\lim_{k \to \infty} W_{k,k} = \infty$ the last sum goes to 0. This gives the second statement in (3.43). The first statement is obvious. $\qquad\square$

Let $\mathcal{W} = \{ W_{j,k}, \, j, k \in \overline{\mathbb{N}} \}$ where

$$W_{j,k} = e^{-|v_k - v_j|}, \qquad \forall \, j, k \in \overline{\mathbb{N}}, \tag{3.46}$$

as defined in (1.47). The next theorem applies Theorem 1.7 when W is written in this way.

Set

$$\overline{\mathcal{K}}_v(j) = \log \left(\sum_{i=1}^{j-1} 1 \wedge 2(v_{i+1} - v_i) \right). \tag{3.47}$$

Theorem 3.1 *Let $\mathcal{W}$ be the potential of a continuous time Markov chain $\mathcal{Z}$ as given in (3.46) and let f be a finite excessive function for $\mathcal{Z}$. Let $\widetilde{\mathcal{Z}}_\alpha = \{ \widetilde{\mathcal{Z}}_{\alpha, j}, \, j \in \overline{\mathbb{N}} \}$ be the α-permanental process with kernel $\widetilde{\mathcal{W}} = \{ \widetilde{W}_{j,k}; \, j, k \in \overline{\mathbb{N}} \}$ where,*

$$\widetilde{W}_{j,k} = W_{j,k} + f_k, \qquad j, k \in \overline{\mathbb{N}}. \tag{3.48}$$

(i) *If $f = \mathcal{W}h$, where $h \in \ell_1^+$, then*

$$\limsup_{j \to \infty} \frac{\widetilde{\mathcal{Z}}_{\alpha, j}}{\overline{\mathcal{K}}_v((j)} = 1, \qquad a.s., \qquad \forall \, \alpha > 0, \tag{3.49}$$

(ii) *If $f = \mathcal{W}h$, where $h \in \ell_1^+$ and $\limsup_{j \to \infty}(v_j - v_{j-1}) < \infty$, then*

$$\limsup_{j\to\infty} \frac{\widetilde{Z}_{\alpha,j}}{\log v_j} = 1, \qquad a.s., \qquad \forall\, \alpha > 0. \tag{3.50}$$

(iii) If $\liminf_{j\to\infty}(v_j - v_{j-1}) > 0$ *and* $f \in c_0^+$, *then*

$$\limsup_{j\to\infty} \frac{\widetilde{Z}_{\alpha,j}}{\log j} = 1, \qquad a.s., \qquad \forall\, \alpha > 0. \tag{3.51}$$

Furthermore when $\liminf_{j\to\infty}(v_j - v_{j-1}) > 0$, *the conditions* $f \in c_0^+$, *and* $f = \widetilde{\mathcal{W}} h$, $h \in c_0^+$, *are equivalent.*

Proof of Theorem 3.1 When $f = \mathcal{W} h$, where $h \in \ell_1^+$, this is simply an application of Theorem 1.7 with s_j replaced by e^{2v_j} and $b_j = s_j^{1/2}$.

That (3.51) extends to α-permanental processes $\widetilde{Z}_\alpha$ with kernels $\widetilde{\mathcal{W}}$ in which f an excessive function for Z with the property that $f \in c_0^+$ follows from Theorem 1.3 since $\liminf_{j\to\infty}(v_j - v_{j-1}) > 0$ implies $\|\mathcal{W}\| < \infty$ and $\mathcal{W}_{j,j} = 1$ for all $j \in \mathbb{N}$.

The fact that $f \in c_0^+$ if and only if $f = \mathcal{W} h$, where $h \in c_0^+$ follows from Lemma 7.2 once we show that (7.20) holds. The condition $\liminf_{j\to\infty}(v_j - v_{j-1}) > 0$ implies that there exists a j_0 such that $(v_j - v_{j-1}) \geq \delta > 0$ for all $j \geq j_0 + 1$. Therefore, for $j \geq 2j_0$,

$$\sum_{k=1}^{j/2} \mathcal{W}_{j,k} \leq \sum_{k=1}^{j/2} e^{-(v_j - v_{j/2})} \leq \frac{j}{2} \delta^{j/2}. \tag{3.52}$$

This shows that (7.20) holds.   □

Clearly, the complete statement involving (3.51) follows from Theorem 1.3 and Lemma 7.2. One doesn't need the much more complicated Theorem 1.2.

Example 3.1 Consider the special case of Theorem 3.1 *(iii)* in which $v_j = \lambda j$, $\lambda > 0$, and $r = e^{-\lambda}$ so that,

$$\widehat{\mathcal{W}}_{j,k} = e^{-\lambda|k-j|} = r^{|k-j|}, \qquad j, k \in \overline{\mathbb{N}}. \tag{3.53}$$

By Lemma 3.2, $\widehat{\mathcal{W}}$ is the potential for the continuous symmetric transient Markov chain on $\overline{\mathbb{N}}$ with Q matrix $D_b Q(s) D_b$, where $b_j = s_j^{1/2} = r^{-(j-1)}$.

This example also follows from Lemma 5.4. We claim that $\widehat{\mathcal{W}}$ is the potential of a continuous symmetric transient Markov chain on $\overline{\mathbb{N}}$ with Q matrix

$$-Q = \frac{1}{1-r^2} \begin{pmatrix} 1 & -r & 0 & 0 & \cdots \\ -r & 1+r^2 & -r & 0 & \cdots \\ 0 & -r & 1+r^2 & -r & \cdots \\ \vdots & \vdots & \vdots & \vdots & \ddots \end{pmatrix}. \tag{3.54}$$

To see this write out

$$\widehat{W} = \begin{pmatrix} 1 & r & r^2 & r^3 & \cdots \\ r & 1 & r & r^2 & \cdots \\ r^2 & r & 1 & r & \cdots \\ \vdots & \vdots & \vdots & \vdots & \ddots \end{pmatrix}. \tag{3.55}$$

It is easily seen that $\widehat{W}Q = -I$.

Consider the birth and death process studied in Chap. 2 which is defined in terms of a strictly increasing sequence $\mathbf{s} = \{s_j, \ j \geq 1\}$ with $s_1 > 0$ and $\lim_{j \to \infty} s_j = \infty$. This process has potential

$$V_{j,k} = s_j \wedge s_k, \qquad j, k \in \overline{\mathbb{N}}. \tag{3.56}$$

We shift the sequence $\mathbf{s}$ by a constant $\Delta > -s_1$ and obtain a new birth and death process defined by the sequence $\mathbf{s}' = \{s'_j = s_j + \Delta, \ j \in \overline{\mathbb{N}}\}$ with potential

$$V'_{j,k} = s'_j \wedge s'_k, \qquad j, k \in \overline{\mathbb{N}}. \tag{3.57}$$

Lemma 3.6 *Let $f = Vh$ for some $h \in \ell_1^+$, then $f = V'h'$ for some $h' \in \ell_1^+$ if*

$$\frac{f_1}{s_1 + \Delta} \geq \frac{f_2 - f_1}{s_2 - s_1}. \tag{3.58}$$

Proof Since $s_j - s_{j-1} = s'_j - s'_{j-1}, j \geq 2$,

$$\frac{f_2 - f_1}{s'_j - s'_{j-1}} = \frac{f_2 - f_1}{s_j - s_{j-1}} \qquad j \geq 2. \tag{3.59}$$

By Corollary 2.1 the right-hand of (3.59) and consequently the left-hand of (3.59) is decreasing. Therefore by Corollary 2.1 again for $f = V'h'$ for some $h' \in \ell_1^+$ we only need in addition that,

$$\frac{f_1}{s'_1} \geq \frac{f_2 - f_1}{s'_2 - s'_1}. \tag{3.60}$$

This is (3.58). $\square$

The next lemma generalizes Lemma 3.2,

Lemma 3.7 *If $b = Vh$ for some $h \in \ell_1^+$ and $b_2 > b_1$ and,*

$$\Delta \leq \frac{b_1 s_2 - b_2 s_1}{b_2 - b_1}, \tag{3.61}$$

then $W' = \{W'_{j,k}, j, k \in \overline{\mathbb{N}}\}$ *where,*

$$W'_{j,k} = \frac{1}{b_j} V'_{j,k} \frac{1}{b_k} = W_{j,k} + \frac{\Delta}{b_j b_k}, \tag{3.62}$$

is the potential of a Markov chain.

Proof Since $b = Vh$ for some $h \in \ell_1^+$ then it follows that if (3.58) holds with f replaced by b, that is if we have,

$$\frac{b_1}{s_1 + \Delta} \geq \frac{b_2 - b_1}{s_2 - s_1}, \tag{3.63}$$

then $b = V'h'$ for some $h' \in \ell_1^+$ in which case (3.62) follows from Lemma 3.2. The condition in (3.61) is simply a rearrangement of (3.63).   □

We see from (2.1) and (2.2) that $Q(s')$ differs from $Q(s)$ only in the $(1, 1)$ entries which are,

$$Q(s)_{1,1} = -\frac{1}{s_1} - Q(s)_{1,2} \quad \text{and} \quad Q(s')_{1,1} = -\frac{1}{s_1 + \Delta} - Q(s)_{1,2}. \tag{3.64}$$

Set

$$\mathcal{Q}(b, s) = D_b Q(s) D_b. \tag{3.65}$$

This is the Q matrix for W. Consequently, $\mathcal{Q}(b, s')$ is the Q matrix for W'. Since $s_j - s_{j-1} = s'_j - s'_{j-1}$ for all $j \geq 2$ and b is unchanged we see that $\mathcal{Q}(b, s')$ differs from $\mathcal{Q}(b, s)$ only in the $(1, 1)$ entry. Using (3.64) and the fact that $\mathcal{Q}(b, s)_{1,2} = b_1 b_2 Q(s)_{1,2}$ we have,

$$\mathcal{Q}(b, s)_{1,1} = -\frac{b_1^2}{s_1} - \frac{b_1}{b_2} \mathcal{Q}(b, s)_{1,2} \quad \text{and} \quad \mathcal{Q}(b, s')_{1,1} = -\frac{b_1^2}{s_1 + \Delta} - \frac{b_1}{b_2} \mathcal{Q}(b, s)_{1,2}. \tag{3.66}$$

Since W' is a potential we know that $\mathcal{Q}(b, s')$ is a Q matrix. Therefore the row sum of its first row must be less than or equal to 0. That is we must have,

$$\mathcal{Q}(b, s')_{1,1} \leq -\mathcal{Q}(b, s')_{1,2} = -\mathcal{Q}(b, s)_{1,2}. \tag{3.67}$$

Using (3.66) and the fact that $\mathcal{Q}(b, s)_{1,2} = b_1 b_2 / (s_2 - s_1)$ we see that this inequality is the same as (3.63).

Example 3.2   Here are some concrete examples of the relationship between $\mathcal{Q}(b, s)$ and W and $\mathcal{Q}(b, s')$ and W'. We take for $\mathcal{Q}(b, s)$ and W the matrices in (3.54) and (3.55). In this case we have $b_j = s_j^{1/2} = r^{-(j-1)}$, $j \geq 1$, and $W_{j,k} = r^{|k-j|}$, $j, k \geq 1$. Therefore, $b_1 = s_1 = 1$, $b_2 = r^{-1}$, $s_2 = r^{-2}$ and,

$$Q(b, \mathbf{s}')_{1,1} = -\frac{1 + r^2 \Delta}{(1 + \Delta)(1 - r^2)}. \tag{3.68}$$

Using the fact that $\Delta > -s_1$ and (3.61) we see that we must have,

$$-1 < \Delta \leq \frac{1}{r}, \quad \text{or equivalently,} \quad Q(b, \mathbf{s}')_{1,1} \leq -\frac{r}{1 - r^2}. \tag{3.69}$$

(i) For $p > 0$ set,

$$\Delta = -r^p \quad \text{or equivalently} \quad Q(b, \mathbf{s}')_{1,1} = -\frac{1 - r^{p+2}}{(1 - r^p)(1 - r^2)}. \tag{3.70}$$

Then by (3.62),

$$W'_{j,k} = r^{|k-j|} - r^{j+k+p-2}. \tag{3.71}$$

(ii) For $p \geq -1$ set,

$$\Delta = r^p \quad \text{or equivalently} \quad Q(b, \mathbf{s}')_{1,1} = -\frac{1 + r^{p+2}}{(1 + r^p)(1 - r^2)}. \tag{3.72}$$

Then by (3.62),

$$W'_{j,k} = r^{|k-j|} + r^{j+k+p-2}. \tag{3.73}$$

(iii) More generally for $\beta \geq r - r^2$ take,

$$\Delta = \frac{1 - \beta - r^2}{\beta} \quad \text{or equivalently} \quad Q(b, \mathbf{s}')_{1,1} = -\frac{\beta + r^2}{1 - r^2}. \tag{3.74}$$

Then by (3.62),

$$W'_{j,k} = r^{|k-j|} + \frac{1 - \beta - r^2}{\beta} r^{j+k-2} \tag{3.75}$$

$$= r^{|k-j|} - \frac{r^{j+k}}{\beta} + \frac{1 - \beta}{\beta} r^{j+k-2}.$$

Remark 3.5 Let $f = W'h$ for some $h \in \ell_1^+$. Let $\widetilde{Z}'_\alpha = \{\widetilde{Z}'_{\alpha,j}, j \in \overline{\mathbb{N}}\}$ be an α-permanental sequence with kernel $\widetilde{W}' = \{\widetilde{W}'_{j,k}; j, k \in \overline{\mathbb{N}}\}$, where

$$\widetilde{W}'_{j,k} = W'_{j,k} + f_k, \quad j, k \in \overline{\mathbb{N}}. \tag{3.76}$$

Then (1.43)–(1.45) hold with $\widetilde{Z}_\alpha$ replaced by $\widetilde{Z}'_\alpha$, and $W_{j,j}$ replaced by $W'_{j,j}$.

This is easy to see. It follows from Theorem 1.7 itself that (1.43)–(1.45) hold with $\widetilde{Z}_\alpha$ replaced by $\widetilde{Z}'_\alpha$ and $W_{j,j}$ replaced by $W'_{j,j}$ and $\mathbf{s}$ replaced by $\mathbf{s}'$. Since $\lim_{j \to \infty} s_j = \infty$ implies that $s'_j \asymp s_j$, $\mathcal{K}_{\mathbf{s}}(j) \asymp \mathcal{K}_{\mathbf{s}'}(j)$ we see that (1.43)–(1.45) hold with $\widetilde{Z}_\alpha$ replaced by $\widetilde{Z}'_\alpha$, and $W_{j,j}$ replaced by $W'_{j,j}$ as stated.

Chapter 4
Birth and Death Processes with Emigration Related to First Order Gaussian Autoregressive Sequences

Let $g_1, g_2, \ldots$ be a sequence of independent identically distributed standard normal random variables and $\{x_n\}$ a sequence of positive numbers. A first order autoregressive Gaussian sequence $\widehat{\xi} = \{\widehat{\xi}_n\}$ is defined by,

$$\widehat{\xi}_1 = g_1, \qquad \widehat{\xi}_n = x_{n-1}\widehat{\xi}_{n-1} + g_n, \qquad n \geq 2. \tag{4.1}$$

It is easy to see that

$$\widehat{\xi}_n = \sum_{i=1}^{n} \left(\prod_{l=i}^{n-1} x_l \right) g_i, \tag{4.2}$$

in which we take the empty product $\prod_{l=n}^{n-1} x_l = 1$.

We consider these processes with the added assumptions that $0 < x_n < 1$, and $x_n \uparrow$.

Let $\mathcal{U} = \{\mathcal{U}_{j,k}; \ j, k \in \overline{\mathbb{N}}\}$ be the covariance matrix for $\widehat{\xi}$. It follows that for $j \leq k$,

$$\mathcal{U}_{j,k} = \sum_{i=1}^{j} \left(\prod_{l=i}^{j-1} x_l \prod_{l=i}^{k-1} x_l \right) = \sum_{i=1}^{j} \left(\prod_{l=i}^{j-1} x_l^2 \prod_{l=j}^{k-1} x_l \right) = \mathcal{U}_{j,j} \prod_{l=j}^{k-1} x_l. \tag{4.3}$$

For $j > k$ we use the fact that $\mathcal{U}$ is symmetric.

We now show that $\mathcal{U}$ can be written in the form of (1.41).

Lemma 4.1

$$\mathcal{U}_{j,k} = \frac{s_j \wedge s_k}{b_j b_k}, \qquad j, k \in \overline{\mathbb{N}}, \tag{4.4}$$

M. B. Marcus and J. Rosen, *Asymptotic Properties of Permanental Sequences*, SpringerBriefs in Probability and Mathematical Statistics, https://doi.org/10.1007/978-3-030-69485-2_4

where,

$$b_j = \prod_{l=1}^{j-1} x_l^{-1}, \quad and \quad s_j = \sum_{i=1}^{j} b_i^2, \quad \forall j \in \overline{\mathbb{N}}. \tag{4.5}$$

Furthermore, $\{s_j\}$ is a strictly increasing convex function of j. (In particular, $\lim_{j \to \infty} s_j = \infty$.)

Proof By (4.3) we have,

$$\mathcal{U}_{j,j} = \sum_{i=1}^{j} \prod_{l=i}^{j-1} x_l^2 = \sum_{i=1}^{j} \prod_{k=1}^{i-1} x_k^{-2} \prod_{l=1}^{j-1} x_l^2 \tag{4.6}$$

$$= \frac{\sum_{i=1}^{j} b_i^2}{b_j^2} = \frac{s_j}{b_j^2}.$$

Using this and (4.3) again, we see that for $j \leq k$,

$$\mathcal{U}_{j,k} = \mathcal{U}_{j,j} \prod_{l=j}^{k-1} x_l = \frac{\mathcal{U}_{j,j}}{\prod_{l=j}^{k-1} x_l^{-1}} \tag{4.7}$$

$$= \frac{\mathcal{U}_{j,j} \prod_{l=1}^{j-1} x_l^{-1}}{\prod_{l=1}^{k-1} x_l^{-1}} = \frac{\mathcal{U}_{j,j} \prod_{l=1}^{j-1} x_l^{-2}}{\prod_{l=1}^{j-1} x_l^{-1} \prod_{l=1}^{k-1} x_l^{-1}} = \frac{\mathcal{U}_{j,j} b_j^2}{b_j b_k} = \frac{s_j}{b_j b_k},$$

which is (4.4).

Since $b_j > 1$, for all $j > 1$, we see that $s_{j+1} - s_j > 1$, for all $j \in \overline{\mathbb{N}}$, so that $s_j \uparrow \infty$.

We can say more than this. By (4.5),

$$(s_{j+1} - s_j) - (s_j - s_{j-1}) = b_{j+1}^2 - b_j^2 > 0, \tag{4.8}$$

which shows that $\{s_j\}$ is an increasing convex function of j. $\qquad \square$

Lemma 4.2 *Let $b = \{b_j\}$, $j \in \overline{\mathbb{N}}$, be as given in (4.5). Then,*

$$b = \mathcal{U}h, \quad for \ some \ h \in \ell_1^+, \tag{4.9}$$

and $\mathcal{U}$ is the potential of a continuous symmetric transient Markov chain on $\overline{\mathbb{N}}$.

Furthermore, the function $g(s_j) = b_j$, $j \in \overline{\mathbb{N}}$, $g(0) = 0$, is an increasing concave function of $\{s_j\}$ and $s_j/b_j \uparrow \infty$.

Proof For $j \geq 2$ we have,

$$\frac{b_j - b_{j-1}}{s_j - s_{j-1}} = \frac{b_j - b_{j-1}}{b_j^2} = \frac{1}{b_j}\left(1 - \frac{b_{j-1}}{b_j}\right) = \frac{1 - x_{j-1}}{b_j}, \tag{4.10}$$

which is decreasing in j. In particular

$$\frac{b_2 - b_1}{s_2 - s_1} = \frac{1 - x_1}{b_2} = x_1(1 - x_1) < 1. \tag{4.11}$$

Therefore, since $b_1 = s_1 = 1$, this shows that $g(s_j) = b_j$ is a concave function on $\{0\} \cup \{s_j, j \geq 1\}$. It follows from Corollary 2.1 that $b = \mathcal{U}h$ for some $h \in \ell_1^+$. Using this and Lemma 3.2 it follows that $\mathcal{U}$ is the potential of a continuous symmetric transient Markov chain on $\overline{\mathbb{N}}$.

It is easy to see that

$$\frac{s_j}{b_j} \uparrow \infty. \tag{4.12}$$

We use the fact that $b_j \uparrow$, which implies that $\lim_{j \to \infty} b_j$ exists. If the limit is finite, (4.12) is trivial because $s_j \to \infty$.

If $\lim_{j \to \infty} b_j = \infty$, (4.12) follows because,

$$\frac{s_j}{b_j} = \mathcal{U}_{j,j} b_j > b_j, \tag{4.13}$$

since $\mathcal{U}_{j,j} > 1$, (see (4.14)).   □

Although it is not obvious the next lemma shows that $\mathcal{U}_{j,j}$ is strictly increasing.

Lemma 4.3 *The terms $\mathcal{U}_{j,j}$ are strictly increasing. Consequently, $\lim_{j \to \infty} \mathcal{U}_{j,j}$ exists. Furthermore, for all $j \geq 2$,*

$$\mathcal{U}_{j,j} \geq 1 + x_{j-1}^2. \tag{4.14}$$

Proof By (4.3),

$$\mathcal{U}_{j,j} = \sum_{i=1}^{j} \prod_{l=i}^{j-1} x_l^2 = \sum_{p=0}^{j-1} \prod_{l=j-p}^{j-1} x_l^2. \tag{4.15}$$

Similarly,

$$\mathcal{U}_{j+1,j+1} = \sum_{p=0}^{j} \prod_{l=j+1-p}^{j} x_l^2 = \prod_{l=1}^{j} x_l^2 + \sum_{p=0}^{j-1} \prod_{l=j+1-p}^{j} x_l^2.$$

Since $x_l \uparrow$,

$$\prod_{l=j+1-p}^{j} x_l^2 \geq \prod_{l=j-p}^{j-1} x_l^2. \tag{4.16}$$

Therefore,

$$\mathcal{U}_{j+1,j+1} \geq \mathcal{U}_{j,j} + \prod_{l=1}^{j} x_l^2. \tag{4.17}$$

This shows that $\{\mathcal{U}_{j,j}\}$ is strictly increasing.

To obtain (4.14) we simply note that for $j \geq 2$, by the first equality in (4.15)

$$\mathcal{U}_{j,j} \geq \prod_{l=j}^{j-1} x_l^2 + \prod_{l=j-1}^{j-1} x_l^2 = 1 + x_{j-1}^2. \tag{4.18}$$

□

Remark 4.1 In Lemma 4.2 we saw that $g(s_j) = b_j$ is an increasing concave function of $\{s_j\}$ and $s_j/b_j \uparrow \infty$. Since $s_j/b_j^2 = \mathcal{U}_{j,j}$, Lemma 4.3 strengthens this to $s_j/b_j^2 \uparrow$. Although it is possible that $\lim_{j \to \infty} s_j/b_j^2 < \infty$.

Lemma 4.4

$$\lim_{j \to \infty} x_j = \delta < 1 \quad \textit{if and only if} \quad \lim_{j \to \infty} \mathcal{U}_{j,j} = \frac{1}{1 - \delta^2}. \tag{4.19}$$

and

$$\lim_{j \to \infty} x_j = 1 \quad \textit{if and only if} \quad \lim_{j \to \infty} \mathcal{U}_{j,j} = \infty, \tag{4.20}$$

Proof Suppose $\sup_j \mathcal{U}_{j,j} < \infty$. Then by Lemma 4.3, $\lim_{j \to \infty} \mathcal{U}_{j,j} = c$, for some $c > 1$. Note that by (4.1),

$$\mathcal{U}_{j+1,j+1} = x_j^2 \mathcal{U}_{j,j} + 1. \tag{4.21}$$

It follows from this that

$$\lim_{j \to \infty} x_j = \left(\frac{c-1}{c} \right)^{1/2}. \tag{4.22}$$

Setting this last expression equal to δ shows that if $\lim_{j \to \infty} \mathcal{U}_{j,j} = 1/1 - \delta^2$, for $0 < \delta < 1$, then $\lim_{j \to \infty} x_j = \delta < 1$.

Now suppose that $\lim_{j \to \infty} x_j = \delta < 1$. Then

$$\mathcal{U}_{j,j} \leq \sum_{i=1}^{j} \delta^{2(i-1)} \leq \frac{1}{1 - \delta^2}. \tag{4.23}$$

This show that $\lim_{j \to \infty} \mathcal{U}_{j,j} = d$, for some $d < \infty$. Taking the limit as $j \to \infty$ in (4.21) we see that

$$d = \delta^2 d + 1. \tag{4.24}$$

Solving for d we see that $\lim_{j \to \infty} \mathcal{U}_{j,j} = 1/(1 - \delta^2)$.

The statement in (4.20) is implied by (4.19).

□

Proof of Theorem 1.8 When $f = \mathcal{U}h$, where $h \in \ell_1^+$ it follows immediately from Theorem 1.7 that if $\limsup_{j\to\infty} s_j/s_{j-1} < \infty$, then,

$$\limsup_{j\to\infty} \frac{\tilde{X}_{\alpha,j}}{\mathcal{U}_{j,j} \log\log(\mathcal{U}_{j,j} b_j^2)} = 1 \quad a.s., \tag{4.25}$$

and if $\liminf_{j\to\infty} s_j/s_{j-1} > 1$, then,

$$\limsup_{j\to\infty} \frac{\tilde{X}_{\alpha,j}}{\mathcal{U}_{j,j} \log j} = 1, \quad a.s., \quad \forall \alpha > 0. \tag{4.26}$$

It follows from (4.21) and (4.6) that,

$$\frac{s_{j+1}}{s_j} = \frac{\mathcal{U}_{j+1,j+1}}{\mathcal{U}_{j,j} x_j^2} = 1 + \frac{1}{x_j^2 \mathcal{U}_{j,j}}. \tag{4.27}$$

We know from Lemma 4.3 that $\lim_{j\to\infty} \mathcal{U}_{j,j}$ exists. If $\lim_{j\to\infty} \mathcal{U}_{j,j} = \infty$, $\lim_{j\to\infty} s_{j+1}/s_j = 1$, which gives (4.25) and (1.52).

If $\limsup_{j\to\infty} \mathcal{U}_{j,j} = c < \infty$, then

$$\liminf_{j\to\infty} s_{j+1}/s_j \geq 1 + \frac{1}{c}, \tag{4.28}$$

which gives (4.26) and, by (4.19), also (1.54).

The proof of (1.53) is given in Lemma 4.8 below.

That (1.54) extends to α-permanental processes $\tilde{X}_\alpha$ with kernels $\tilde{\mathcal{U}}$ in which f an excessive function for $\mathcal{X}$ with the property that $f \in c_0^+$, follows from Theorem 1.3. We show that the conditions in (1.20) are satisfied. Since

$$\mathcal{U}_{j,k} = \sum_{i=1}^{j} \left(\prod_{l=i}^{j-1} x_l \prod_{l=i}^{k-1} x_l \right), \quad j \leq k, \tag{4.29}$$

and $\{x_j\}$ is an increasing sequence, we have

$$\mathcal{U}_{j,k} \leq \sum_{i=1}^{j} \delta^{(j-i)+(k-i)} = \delta^{j+k} \sum_{i=1}^{j} \delta^{-2i} \leq \delta^{k-j} \frac{1}{1-\delta^2}, \quad j \leq k. \tag{4.30}$$

Therefore,

$$\sum_{k=j}^{\infty} \mathcal{U}_{j,k} \leq \frac{1}{(1-\delta)(1-\delta^2)}. \tag{4.31}$$

Since this also holds when $k \le j$ we see that $\|\mathcal{U}\| < \infty$. In addition it follows from (4.29) that $\mathcal{U}_{j,j} \ge 1$.

The fact that $f \in c_0^+$ if and only if $f = \mathcal{U}h$, where $h \in c_0^+$ follows from Lemma 7.2 once we show that (7.20) holds. This is easy to see since,

$$\sum_{j=1}^{k/2} \mathcal{U}_{j,k} \le \frac{\delta^{k/2}}{(1-\delta^2)} \sum_{j=1}^{k/2} \delta^{k/2-j} \le \frac{\delta^{k/2}}{(1-\delta)(1-\delta^2)}. \tag{4.32}$$

$\square$

We use the next lemma to obtain Example 1.1 (i) and (ii).

Lemma 4.5 *If $\lim_{j\to\infty} j(1 - x_j^2) = c$, for some $c \ge 0$, then*

$$\mathcal{U}_{j,j} \sim \frac{1}{1+c}\, j \quad as \quad j \to \infty, \tag{4.33}$$

and

$$\log\log(\mathcal{U}_{j,j} b_j^2) \sim \log\log j \quad as \ j \to \infty. \tag{4.34}$$

Proof We have,

$$\mathcal{U}_{j,j} = \sum_{i=1}^{j} \left(\prod_{l=i}^{j-1} x_l^2 \right). \tag{4.35}$$

For some $\epsilon > 0$ let $0 < a < c < b$ be such that $|a - b| < \epsilon$. Since $\lim_{l\to\infty} l(1 - x_l^2) = c$ we can find a j_0 such that,

$$-\frac{b}{l} < \log x_l^2 < -\frac{a}{l}, \qquad \forall l \ge j_0. \tag{4.36}$$

Consequently,

$$\sum_{i=j_0}^{j} \prod_{l=i}^{j-1} x_l^2 = \sum_{i=j_0}^{j} \exp\left(\sum_{l=i}^{j-1} \log x_l^2 \right) \le \sum_{i=j_0}^{j} \exp\left(-\sum_{l=i}^{j-1} \frac{a}{l} \right) \tag{4.37}$$

$$\sim \frac{1}{(j-1)^a} \sum_{i=j_0}^{j} i^a \sim \frac{j}{1+a}, \qquad as \ j \to \infty.$$

Similarly,

$$\sum_{i=j_0}^{j} \prod_{l=i}^{j-1} x_l^2 \ge \frac{j}{1+b}, \qquad as \ j \to \infty. \tag{4.38}$$

Note that the left-hand sides of (4.37) and (4.38) differ from (4.35) by some finite number. Therefore, since (4.36) holds for all $\epsilon > 0$, we get (4.33) when $c > 0$.

When $c = 0$ the left-hand side of (4.36) holds for all $b > 0$. Therefore,

$$\liminf_{j \to \infty} \frac{\mathcal{U}_{j,j}}{j} \geq 1. \tag{4.39}$$

However, since $\mathcal{U}_{j,j} \leq j$, we get (4.33) when $c = 0$.

To get (4.34) we note that by (4.33),

$$\log \log(\mathcal{U}_{j,j} b_j^2) \sim \log \left(\log \left(\frac{j}{1+c} \right) + \log b_j^2 \right), \qquad \text{as } j \to \infty. \tag{4.40}$$

Furthermore, by (4.33)

$$\log b_j^2 = \log \prod_{l=1}^{j-1} x_l^{-2} = -\sum_{l=1}^{j-1} \log x_l^2 \tag{4.41}$$

$$= C - \sum_{l=j_0}^{j-1} \log x_l^2 \leq C + \sum_{l=j_0}^{j-1} \frac{b}{l} \leq C + b \log j,$$

where $C = -\sum_{l=1}^{j_0-1} \log x_l^2$. Using (4.40) and (4.41) and the fact that $\log b_j \geq 0$, we get (4.34). □

Proof of Example 1.1 (*i*) This follows immediately from Lemma 4.5. We now show that this includes the case where $\prod_{j=1}^{\infty} x_j > 0$. Set $x_j = 1 - \epsilon_j$. Therefore, for some $C > 0$,

$$\prod_{j=1}^{\infty} x_j \leq C \exp \left(-\sum_{j=1}^{\infty} \epsilon_j \right). \tag{4.42}$$

If $\prod_{j=1}^{\infty} x_j > 0$ then $\sum_{j=1}^{\infty} \epsilon_j < \infty$. Since ϵ_j is decreasing it follows from Lemma 4.6 that $\epsilon_j = o(1/j)$, as $j \to \infty$. Therefore the condition that $j(1 - x_j^2) \to 0$, as $j \to \infty$ includes the case when $\prod_{j=1}^{\infty} x_j > 0$. □

Lemma 4.6 *If $\epsilon_j \downarrow$ and $\sum_{j=1}^{\infty} \epsilon_j < \infty$, then $\epsilon_j = o(1/j)$, as $j \to \infty$.*

Proof Suppose that $\limsup_{j \to \infty} j\epsilon_j \geq \delta > 0$. Then we can find a subsequence $\{j_k\}$ such that

$$\frac{j_k}{j_{k+1}} \leq \frac{1}{2}, \qquad \text{and} \qquad \epsilon_{j_k} \geq \frac{\delta}{j_k}. \tag{4.43}$$

Therefore,

$$\sum_{j=1}^{\infty} \epsilon_j = \sum_{k=1}^{\infty} \sum_{l=j_k}^{j_{k+1}-1} \epsilon_j \geq \sum_{k=1}^{\infty} \epsilon_{j_{k+1}} (j_{k+1} - j_k) \tag{4.44}$$

$$\geq \delta \sum_{k=1}^{\infty} \left(1 - \frac{j_k}{j_{k+1}} \right) = \infty.$$

$\square$

Proof of Example 1.1 (*ii*) This follows immediately from Lemma 4.5. $\square$

The next lemma gives some useful information about $\mathcal{U}_{j,j}$:

Lemma 4.7

$$\mathcal{U}_{j,j} \leq \frac{1}{1 - x_j^2}. \tag{4.45}$$

Furthermore the following are equivalent:

$$\mathcal{U}_{j,j} \sim \frac{1}{1 - x_j^2}, \quad \text{as} \;\; j \to \infty, \tag{4.46}$$

and

$$\mathcal{U}_{j+1,j+1} - \mathcal{U}_{j,j} \to 0, \quad \text{as} \; j \to \infty. \tag{4.47}$$

Proof By (4.21) and Lemma 4.3,

$$(1 - x_j^2)\mathcal{U}_{j,j} = 1 - (\mathcal{U}_{j+1,j+1} - \mathcal{U}_{j,j}) \leq 1. \tag{4.48}$$

All the statements in this lemma follow easily from this. $\square$

Using Lemma 4.7 we make (1.52) more specific:

Lemma 4.8 *In Theorem 1.8 assume that $\mathcal{U}_{j,j}$ is a regularly varying function with index $0 < \beta < 1$ then,*

$$\limsup_{j \to \infty} \frac{\tilde{\mathcal{X}}_{\alpha,j}}{\mathcal{U}_{j,j} \log j} = 1 - \beta, \quad a.s. \quad \forall \alpha > 0. \tag{4.49}$$

Proof Considering (1.52) we need to show that

$$\log \log(\mathcal{U}_{j,j} b_j^2) = \log(\log \mathcal{U}_{j,j} + \log b_j^2) \sim (1 - \beta) \log j, \quad \text{as} \; j \to \infty. \tag{4.50}$$

Furthermore, since $\mathcal{U}_{j,j} \leq j$, we need to show that,

$$\log \log b_j^2 \sim (1 - \beta) \log j, \quad \text{as} \; j \to \infty. \tag{4.51}$$

To be specific let $\mathcal{U}_{j,j} = g(j) = j^{\beta} L(j)$, where L is a slowly varying function. Then clearly, (4.47) holds. Therefore, by (4.46), $(1 - x_j^2) \sim 1/g(j)$. It follows that for all $\epsilon > 0$ and $0 < a < 1 < b$ such that $|a - b| \le \epsilon$ we can find an integer j_0 such that

$$-\frac{b}{g(l)} < \log x_l^2 < -\frac{a}{g(l)}, \qquad \forall l \ge j_0. \tag{4.52}$$

Similar to (4.41), for all j sufficiently large,

$$\log b_j^2 = -\sum_{l=1}^{j-1} \log x_l^2 \tag{4.53}$$

$$\le -\sum_{l=1}^{j_0-1} \log x_l^2 + \sum_{l=j_0}^{j-1} \frac{b}{g(l)} \sim \frac{bj^{1-\beta}}{(1-\beta)L(j)}.$$

Likewise,

$$\log b_j^2 \ge -\sum_{l=1}^{j_0-1} \log x_l^2 + \sum_{l=j_0}^{j-1} \frac{a}{g(l)} \sim \frac{aj^{1-\beta}}{(1-\beta)L(j)}.$$

These two inequalities give (4.51). $\qquad\qquad\qquad\qquad\qquad\qquad\qquad\square$

In the proof of Lemma 4.8 we use the fact that when $\mathcal{U}_{j,j}$ is a regularly varying function with index $0 < \beta < 1$, then $(1 - x_j^2) \sim 1/\mathcal{U}_{j,j}$. What we do not show is that when $(1 - x_j^2) \sim h(j)$, for some regularly varying function $h(j)$ with index $-1 < \beta' < 0$ then $\mathcal{U}_{j,j} \sim 1/h(j)$. We only consider this in the special case given in Example 1.1 (iii).

Proof of Example 1.1 (iii) We have,

$$\mathcal{U}_{j,j} = \sum_{i=1}^{j} \left(\prod_{l=i}^{j-1} x_l^2 \right). \tag{4.54}$$

For some $\epsilon > 0$ let $0 < a < 1 < b$ be such that $|a - b| < \epsilon$. Since $\lim_{l \to \infty} l^{\beta}(1 - x_l^2) = 1$ we can find a j_0 such that,

$$-\frac{b}{l^{\beta}} < \log x_l^2 < -\frac{a}{l^{\beta}}, \qquad \forall l \ge j_0. \tag{4.55}$$

Consequently,

$$\sum_{i=j_0}^{j} \prod_{l=i}^{j-1} x_l^2 = \sum_{i=j_0}^{j} \exp\left(\sum_{l=i}^{j-1} \log x_l^2\right) \leq \sum_{i=j_0}^{j} \exp\left(-\sum_{l=i}^{j-1} \frac{a}{l^\beta}\right) \tag{4.56}$$

$$\sim \exp\left(-\frac{aj^{1-\beta}}{1-\beta}\right) \int_{j_0}^{j} \exp\left(\frac{ax^{1-\beta}}{1-\beta}\right) dx, \qquad \text{as } j \to \infty,$$

and

$$\int_{j_0}^{j} \exp\left(\frac{ax^{1-\beta}}{1-\beta}\right) dx = \frac{1}{a} \int_{j_0}^{j} x^\beta d\left(\exp\left(\frac{ax^{1-\beta}}{1-\beta}\right)\right) \tag{4.57}$$

$$\sim \frac{j^\beta}{a} \exp\left(\frac{aj^{1-\beta}}{1-\beta}\right), \qquad \text{as } j \to \infty,$$

where, for the last line we use integration by parts. Therefore,

$$\sum_{i=j_0}^{j} \prod_{l=i}^{j-1} x_l^2 \leq \frac{j^\beta}{a}, \qquad \text{as } j \to \infty. \tag{4.58}$$

A similar argument shows that the left-hand side of (4.58) is greater than or equal to j^β/b as $j \to \infty$. Using these observations and following the proof of Lemma 4.5 we see that

$$\mathcal{U}_{j,j} \sim j^\beta, \qquad \text{as } j \to \infty. \tag{4.59}$$

Therefore (1.57) follows from (1.53).   □

We now explicitly describe the Q matrix corresponding to $\mathcal{U}$ in Lemma 4.1, which is

$$Q(b, s) := D_b Q(s) D_b. \tag{4.60}$$

(See (2.1) and (2.15).)

It follows from (4.5) that,

$$a_j = \frac{1}{s_j - s_{j-1}} = b_j^{-2}, \qquad j \geq 2, \tag{4.61}$$

and

$$a_1 = \frac{1}{s_1} = b_1^{-2} = 1. \tag{4.62}$$

Therefore (4.61) holds for all $j \geq 1$. Consequently, for all $j \geq 1$,

$$-Q(b, s)_{j,j+1} = -b_j a_{j+1} b_{j+1} = -\frac{b_j}{b_{j+1}} = -x_j, \tag{4.63}$$

and

$$- Q(b, s)_{j,j} = b_j \left(a_j + a_{j+1} \right) b_j = b_j^2 \left(\frac{1}{b_j^2} + \frac{1}{b_{j+1}^2} \right) = 1 + x_j^2. \qquad (4.64)$$

Since $Q(b, s)_{j, j+1} = Q(b, s)_{j+1, j}$ we have,

$$- Q(b, s) = \begin{pmatrix} 1 + x_1^2 & -x_1 & 0 & 0 & \cdots & 0 & 0 & \cdots \\ -x_1 & 1 + x_2^2 & -x_2 & 0 & \cdots & 0 & 0 & \cdots \\ 0 & -x_2 & 1 + x_3^2 & -x_3 & \cdots & 0 & 0 & \cdots \\ \vdots & \vdots & \vdots & \vdots & \ddots & \vdots & \vdots & \ddots \\ 0 & 0 & 0 & 0 & \cdots & 1 + x_m^2 & -x_m & \cdots \\ 0 & 0 & 0 & 0 & \cdots & -x_m & 1 + x_{m+1}^2 & \cdots \\ \vdots & \vdots & \vdots & \vdots & \ddots & \vdots & \vdots & \ddots \end{pmatrix} \qquad (4.65)$$

Example 4.1 Let $x_j = r$. Then $b_j = r^{-(j-1)}$ and

$$- Q(s, b) = \begin{pmatrix} 1 + r^2 & -r & 0 & 0 & \cdots \\ -r & 1 + r^2 & -r & 0 & \cdots \\ 0 & -r & 1 + r^2 & -r & \cdots \\ \vdots & \vdots & \vdots & \vdots & \ddots \end{pmatrix}, \qquad (4.66)$$

In addition

$$\mathcal{U}_{j,j} = 1 + r^2 + r^4 + \cdots + r^{2(j-1)} = \frac{1 - r^{2j}}{1 - r^2} \qquad (4.67)$$

and for $j \le k$,

$$\mathcal{U}_{j,k} = \mathcal{U}_{j,j} r^{k-j} = \frac{r^{k-j} - r^{k+j}}{1 - r^2}. \qquad (4.68)$$

Consequently,

$$\mathcal{U}_{j,k} = \mathcal{U}_{j,j} r^{k-j} = \frac{r^{|k-j|} - r^{k+j}}{1 - r^2}, \qquad \forall j, k \in \overline{\mathbb{N}}. \qquad (4.69)$$

Compare (3.66) with $p = 2$. (Note that

$$\frac{e^{-\sqrt{2\delta}|x-y|} - e^{-\sqrt{2\delta}x} e^{-\sqrt{2\delta}y}}{\sqrt{2\delta}}, \qquad (4.70)$$

the δ-potential density for Brownian motion killed the first time it hits 0.)

We show in Lemma 4.2 that the covariance of the first order Gaussian autoregressive sequence $\widehat{\xi}$ in (4.1) is the potential of a continuous time Markov chain

$\mathcal{U} = \{U_{j,k}; j, k \in \overline{\mathbb{N}}\}$ where,

$$U_{j,k} = \frac{s_j \wedge s_k}{b_j b_k}. \tag{4.71}$$

At the end of Chap. 3 we consider the effect of a shift $s_j \to s'_j = s_j + \Delta$ on such potentials. We now show that when apply such a shift to $\mathcal{U}$ we still have the covariance of a first order Gaussian autoregressive sequence.

As in (4.1), let $g_1, g_2, \ldots$ be a sequence of independent identically distributed standard normal random variables, $0 < x_n < 1$, $x_n \uparrow$, and take $\widetilde{\delta} \neq 0$. Consider the Gaussian sequences $\widetilde{\xi} = \{\widetilde{\xi}_n\}$ defined by,

$$\widetilde{\xi}_1 = \widetilde{\delta} g_1, \quad \widetilde{\xi}_n = x_{n-1}\widetilde{\xi}_{n-1} + g_n, \quad n \geq 2. \tag{4.72}$$

This generalizes $\widehat{\xi}$ in (4.1).

Theorem 4.1 *Let $\{s_j\}$ and $\{b_j\}$ be as given in Lemma 4.1. The covariance of the first order Gaussian auto regressive sequence $\widetilde{\xi}$ is $\mathcal{U}' = \{U'_{j,k}; j, k \in \overline{\mathbb{N}}\}$ where,*

$$U'_{j,k} = \frac{s'_j \wedge s'_k}{b_j b_k}, \tag{4.73}$$

$$s'_j = s_j + \Delta, \quad and \quad \Delta = \widetilde{\delta}^2 - 1, \quad 0 < \widetilde{\delta} < \infty. \tag{4.74}$$

Furthermore, $\mathcal{U}'$ is the potential of a transient Markov chain if

$$0 < \widetilde{\delta}^2 \leq \frac{1}{x_1(1 - x_1)}. \tag{4.75}$$

Proof It is easy to see that

$$\widetilde{\xi}_n = \sum_{i=1}^{n} \left(\prod_{l=i}^{n-1} y_l\right) g_i, \quad n \geq 2, \tag{4.76}$$

where $y_1 = \widetilde{\delta} x_1$, $y_l = x_l, l \geq 2$ and in which we take the empty product $\prod_{l=n}^{n-1} y_l = 1$. Therefore, for $j \leq k$,

$$U'_{j,k} = \sum_{i=1}^{j} \left(\prod_{l=i}^{j-1} y_l \prod_{l=i}^{k-1} y_l\right) = \sum_{i=1}^{j} \left(\prod_{l=i}^{j-1} y_l^2 \prod_{l=j}^{k-1} y_l\right) = U'_{j,j} \prod_{l=j}^{k-1} y_l \tag{4.77}$$

$$= \frac{U'_{j,j}}{\prod_{l=j}^{k-1} y_l^{-1}} = \frac{U'_{j,j} \prod_{l=1}^{j-1} y_l^{-1}}{\prod_{l=1}^{k-1} y_l^{-1}} = \frac{U'_{j,j} b_j}{b_k}.$$

Set $s'_j = b_j^2 U'_{j,j}$, $j \in \overline{\mathbb{N}}$, so that (4.73) holds. Also note that,

$$\mathcal{U}'_{j,j} = \sum_{i=1}^{j} \prod_{l=i}^{j-1} y_l^2 = \prod_{l=1}^{j-1} y_l^2 + \sum_{i=2}^{j} \prod_{l=i}^{j-1} y_l^2 \tag{4.78}$$

$$= \widetilde{\delta}^2 \prod_{l=1}^{j-1} x_l^2 + \sum_{i=2}^{j} \prod_{l=i}^{j-1} x_l^2.$$

Therefore, since $b_1 = 1$,

$$s'_j = b_j^2 \mathcal{U}'_{j,j} = \widetilde{\delta}^2 + \sum_{i=2}^{j} \prod_{l=1}^{i-1} x_l^{-2} = \widetilde{\delta}^2 + \sum_{i=2}^{j} b_i^2 = (\widetilde{\delta}^2 - 1) + s_j, \tag{4.79}$$

where we use (4.5) for the last equation. This gives $s'_j = s_j + \Delta$ with $\Delta = \widetilde{\delta}^2 - 1$.

It follows from Lemma 3.7 and (4.5) that when (4.75) holds, $\mathcal{U}'$ is the potential of a Markov chain. □

Remark 4.2 Assume condition (4.75), so that $\mathcal{U}'$ is the potential of a transient Markov chain which we denote by $\mathcal{X}'$. Let f be a finite excessive function for $\mathcal{X}'$. Let $\widetilde{\mathcal{X}}'_\alpha = \{\widetilde{\mathcal{X}}'_{\alpha,j}, j \in \overline{\mathbb{N}}\}$ be an α-permanental sequence with kernel $\widetilde{\mathcal{U}}' = \{\widetilde{\mathcal{U}}'_{j,k}; j, k \in \overline{\mathbb{N}}\}$, where

$$\widetilde{\mathcal{U}}'_{j,k} = \mathcal{U}'_{j,k} + f_k, \qquad j, k \in \overline{\mathbb{N}}. \tag{4.80}$$

Then using the same argument used in Remark 3.5 we see that if $f = \mathcal{U}'h$ for some $h \in \ell_1$ then (1.52) and (1.53) hold with $\widetilde{X}_\alpha$ replaced by $\widetilde{X}'_\alpha$. Item (ii) in Theorem 1.8 also holds with $\widetilde{X}_\alpha$ replaced by $\widetilde{X}'_\alpha$.

Example 1.1 also holds with $\widetilde{X}_\alpha$ replaced by $\widetilde{X}'_\alpha$ since the computations depend on the relationship between $\mathcal{U}$ and $\{b_j\}$ and $\{b_j\}$ is unchanged.

Remark 4.3 Condition (4.75) is necessary for $\mathcal{U}'$ to be the potential of a Markov chain whereas (4.73) holds for all $\widetilde{\delta} \neq 0$. This gives examples of a critical point at which a covariance matrix ceases to be an inverse M-matrix. This has interesting implications in the study of Gaussian sequences with infinitely divisible squares.

Chapter 5
Markov Chains with Potentials That Are the Covariances of Higher Order Gaussian Autoregressive Sequences

Consider a class of k-th order autoregressive Gaussian sequences, for $k \geq 2$. Let $g_1, g_2, \ldots$ independent standard normal random variables and let $p_i > 0$, $i = 1, \ldots, k$, with $\sum_{l=1}^{k} p_l \leq 1$. We define the Gaussian sequence $\widetilde{\xi} = \{\widetilde{\xi}_n, n \in \overline{\mathbb{N}}\}$ by,

$$\widetilde{\xi}_1 = g_1, \quad \text{and} \quad \widetilde{\xi}_n = \sum_{l=1}^{k} p_l \widetilde{\xi}_{n-l} + g_n, \quad n \geq 2, \tag{5.1}$$

where $\widetilde{\xi}_i = 0$ for all $i \leq 0$. Let $\mathcal{V} = \{\mathcal{V}_{j,k}; j, k \in \overline{\mathbb{N}}\}$ denote the covariance of $\widetilde{\xi}$.

Our goal is to prove Theorem 1.9. We begin by exhibiting some simple properties $\mathcal{V}$. Set

$$\phi_n = E(\widetilde{\xi}_n g_1), \tag{5.2}$$

and note that

$$\phi_1 = 1 \quad \text{and} \quad \phi_n = \sum_{l=1}^{k} p_l \phi_{n-l}, \quad n \geq 2, \tag{5.3}$$

where $\phi_n = 0$ for all $n \leq 0$. Since $\phi_2 = p_1 < 1$ and $\sum_{l=1}^{k} p_l \leq 1$, we see that,

$$\phi_n < 1, \quad \forall n \geq 2. \tag{5.4}$$

We now write $\{\widetilde{\xi}_n\}$ as a series with terms that are independent Gaussian random variables.

Lemma 5.1

$$\widetilde{\xi}_n = \sum_{j=1}^{n} \phi_{n+1-j} g_j = \sum_{j} \phi_{n+1-j} g_j, \quad j \in \overline{\mathbb{N}}, \tag{5.5}$$

© The Author(s), under exclusive license to Springer Nature Switzerland AG 2021
M. B. Marcus and J. Rosen, *Asymptotic Properties of Permanental Sequences*,
SpringerBriefs in Probability and Mathematical Statistics,
https://doi.org/10.1007/978-3-030-69485-2_5

(since the terms in the last sum are all equal to 0 when $j \notin [1, n]$). Therefore,

$$\mathcal{V}_{m,n} = E(\tilde{\xi}_m \tilde{\xi}_n) = \sum_{j=1}^{m \wedge n} \phi_{m+1-j} \phi_{n+1-j} = \sum_{j=0}^{(m \wedge n)-1} \phi_{m-j} \phi_{n-j}, \tag{5.6}$$

which implies, in particular, that

$$\mathcal{V}_{m,n} \leq m \wedge n, \quad \mathcal{V}_{1,1} = 1, \quad and \quad \mathcal{V}_{n,n} = E(\tilde{\xi}_n^2) \uparrow. \tag{5.7}$$

Proof We give a proof by induction. Clearly (5.5) is true for $n = 1$. Then using (5.1) and induction we have

$$\tilde{\xi}_n = \sum_{l=1}^{k} p_l \tilde{\xi}_{n-l} + g_n \tag{5.8}$$

$$= \sum_{l=1}^{k} p_l \sum_{j \leq n-1} \phi_{n-l+1-j} g_j + g_n,$$

where in the second equality we change nothing by allowing $j \leq n - 1$ rather than $j \leq n - l$, since $\phi_n = 0$ for $n < 1$. Interchanging the order of summation this is equal to

$$\sum_{j \leq n-1} \left(\sum_{l=1}^{k} p_l \phi_{n-l+1-j} \right) g_j + g_n \tag{5.9}$$

$$= \sum_{j \leq n-1} \phi_{n+1-j} g_j + g_n$$

where the last equality came from (5.3), since for $j \leq n - 1$ we have $n + 1 - j \geq 2$. This gives (5.5).

The statement in (5.6) follows immediately from (5.5); (5.7) is an immediate consequence of (5.6), (5.4) and (5.3), since $\mathcal{V}_{1,1} = \phi_1^2$. □

We now introduce the matrix A which, with the additional condition that its off diagonal elements are less that or equal to 0, is the negative of the Q matrix for the continuous time symmetric Markov chains on $\overline{\mathbb{N}}$ with potential $V = \{\mathcal{V}_{j,k}, j, k \in \overline{\mathbb{N}}\}$.

Lemma 5.2 *Let $A = \{A_{m,n}; m, n \in \overline{\mathbb{N}}\}$ where,*

$$A_{m,m} = 1 + \sum_{i=1}^{k} p_i^2, \quad \forall m \in \mathbb{N}, \tag{5.10}$$

$$A_{m,n} = -p_{|m-n|} + \sum_{l=1}^{k-|m-n|} p_l\, p_{|m-n|+l}, \qquad for\ all\ 1 \le |m-n| \le k, \quad (5.11)$$

and

$$A_{m,n} = 0, \qquad \forall\, |m-n| > k. \quad (5.12)$$

Then

$$\mathcal{V}A = A\mathcal{V} = I, \quad (5.13)$$

in the sense of multiplication of matrices. That is, for each $i, l \in \overline{\mathbb{N}}$,

$$\sum_j \mathcal{V}_{i,j} A_{j,l} = \delta_{i,l}, \quad (5.14)$$

and similarly $A\mathcal{V}$.

Clearly $A_{m,n}$ depends only on $|m-n|$. Set

$$a_{|m-n|} = A_{m,n}, \qquad n, m \in \overline{\mathbb{N}}. \quad (5.15)$$

Note that A is a symmetric Töeplitz matrix and that for $j \ge k+1$, the j-th row of A has the form

$$0, \ldots, 0, a_k, \ldots, a_1, a_0, a_1, \ldots, a_k, 0, 0, \ldots, \quad (5.16)$$

where the initial sequence of zeros has $j - k$ terms.
For $j \le k$ the $(j+1)$-st row of A has the form

$$a_j, \ldots, a_1, a_0, a_1, \ldots, a_k, 0, 0, \ldots. \quad (5.17)$$

Here is an explicit example.

Example 5.1 When $k = 2$,

$$A = \begin{pmatrix} 1+p_1^2+p_2^2 & -p_1+p_1 p_2 & -p_2 & 0 & 0 & \cdots \\ -p_1+p_1 p_2 & 1+p_1^2+p_2^2 & -p_1+p_1 p_2 & -p_2 & 0 & \cdots \\ -p_2 & -p_1+p_1 p_2 & 1+p_1^2+p_2^2 & -p_1+p_1 p_2 & -p_2 & \cdots \\ 0 & -p_2 & -p_1+p_1 p_2 & 1+p_1^2+p_2^2 & -p_1+p_1 p_2 & \cdots \\ \vdots & \vdots & \vdots & \vdots & \vdots & \ddots \end{pmatrix} \quad (5.18)$$

We see that in this case A is a symmetric Töeplitz matrix with five non-zero diagonals. The row sums for all rows after the second row are equal to $(1 - (p_1 + p_2))^2$. Note also that $-A$ is a Q-matrix, since, $p_1 p_2 \le p_1$.

Proof of Lemma 5.2 We introduce two infinite matrices,

$$
L = \begin{pmatrix}
1 & 0 & 0 & \cdots & 0 & 0 & \vdots & \vdots & \cdots \\
-p_1 & 1 & 0 & \cdots & 0 & 0 & \vdots & \vdots & \cdots \\
-p_2 & -p_1 & 1 & \cdots & 0 & 0 & \vdots & \vdots & \cdots \\
\vdots & \vdots & \vdots & \ddots & \vdots & \vdots & \vdots & \vdots & \cdots \\
-p_k & -p_{k-1} & -p_{k-2} & \cdots & -p_1 & 1 & 0\ 0 & \cdots \\
0 & -p_k & -p_{k-1} & \cdots & -p_2 & -p_1 & 1\ 0 & \cdots \\
\vdots & \vdots & \vdots & \ddots & \vdots & \vdots & \vdots\ \vdots & \ddots
\end{pmatrix},
\tag{5.19}
$$

and

$$
\Phi = \begin{pmatrix}
\phi_1 & 0 & 0 & \cdots & 0\ 0\ 0\ 0 & \cdots \\
\phi_2 & \phi_1 & 0 & \cdots & 0\ 0\ 0\ 0 & \cdots \\
\phi_3 & \phi_2 & \phi_1 & \cdots & 0\ 0\ 0\ 0 & \cdots \\
\vdots & \vdots & \vdots & \ddots & \vdots\ \vdots\ \vdots\ \vdots & \ddots \\
\phi_n & \phi_{n-1} & \phi_{n-2} & \cdots & \phi_1\ 0\ 0\ 0 & \cdots \\
\vdots & \vdots & \vdots & \ddots & \vdots\ \vdots\ \vdots\ \vdots & \ddots
\end{pmatrix}.
\tag{5.20}
$$

where $\{\phi_n\}$ is given in (5.3).

It is easy to see that,

$$
L\Phi = \Phi L = I, \quad \text{and} \quad L^T \Phi^T = \Phi^T L^T = I.
\tag{5.21}
$$

We also give is an analytical proof. Set $p_0 = -1$ and $p_j = 0$, $j < 0$, and write,

$$
L_{i,j} = -p_{i-j}, \quad i, j \in \overline{\mathbb{N}}.
\tag{5.22}
$$

and

$$
\Phi_{i,j} = \phi_{i+1-j}, \quad i, j \in \overline{\mathbb{N}}.
\tag{5.23}
$$

Consequently,

$$
(L\Phi)_{m,n} = -\sum_j p_{m-j}\phi_{j+1-n}
\tag{5.24}
$$

$$
= \phi_{m+1-n} - \sum_{n \le j < m} p_{m-j}\phi_{j+1-n}.
$$

When $n = m$ there are no non-zero terms in the final sum in (5.24) and since $\phi_1 = 1$ we have $(L\Phi)_{n,n} = 1$. If $m < n$, all the terms in the last line of (5.24) are equal to 0, so we have $(L\Phi)_{m,n} = 0$. When $m > n$, we set $l = m - j$ and write (5.24) as,

$$(L\Phi)_{m,n} = \phi_{m+1-n} - \sum_{l=1}^{m-n} p_l \phi_{m+1-n-l} = 0, \tag{5.25}$$

which follows from (5.3). Thus we see that $L\Phi = I$. The second equality in (5.21) follows similarly. The last two equalities in (5.21) follow immediately.

We now obtain (5.13). Note that it follows from (5.6) that for all $m, n \in \overline{\mathbb{N}}$,

$$\left(\Phi\Phi^T\right)_{m,n} = \sum_{j=1}^{m\wedge n} \phi_{m+1-j}\phi_{n+1-j} = E\left(\widetilde{\xi}_m \widetilde{\xi}_n\right) = \mathcal{V}_{m,n}. \tag{5.26}$$

We show below that $A = L^T L$. Therefore,

$$\sum_j \mathcal{V}_{i,j} A_{j,l} = \sum_j \left(\sum_m \Phi_{i,m}\Phi_{m,j}^T\right)\left(\sum_n L_{j,n}^T L_{n,l}\right). \tag{5.27}$$

It is easy to see that (5.14) holds, once we show that we can interchange the order of summation in (5.27). This allows us to write,

$$\sum_j \mathcal{V}_{i,j} A_{j,l} = \sum_m \sum_n \Phi_{i,m}\left(\sum_j \Phi_{m,j}^T L_{j,n}^T\right) L_{n,l} \tag{5.28}$$

$$= \sum_m \sum_n \Phi_{i,m}\delta_{m,n} L_{n,l} = \sum_n \Phi_{i,n} L_{n,l} = \delta_{i,l},$$

where we use (5.21) twice.

To show that we can interchange the order of summation in (5.27) it suffices to show that for i and l fixed all the sums in (5.27) are only over a finite number of terms that are not equal to 0. Making use of the fact that many of the terms in L and Φ are equal to 0, we see that,

$$\sum_m \Phi_{i,m}\Phi_{m,j}^T = \sum_{m=1}^{i} \Phi_{i,m}\Phi_{m,j}^T \tag{5.29}$$

and

$$\sum_n L_{j,n}^T L_{n,l} = \sum_{n=l}^{l+k+1} L_{j,n}^T L_{n,l}. \tag{5.30}$$

Furthermore, for each n, $L_{j,n}^T = 0$ when $j > n$. This shows that the summation in (5.27) is only over a finite number of terms.

We show in (5.28) that $\mathcal{V}A = I$. Since both $\mathcal{V}$ and A are symmetric, we also have $A\mathcal{V} = I$.

To show that $A = L^T L$ we take the product $L^T L$ to see that

$$A_{m,m} = (L^T L)_{m,m} = \sum_j L_{j,m}^2 = \sum_j p_{j-m}^2 = 1 + \sum_{i=1}^{k} p_i^2, \tag{5.31}$$

and for $n < m$,

$$(L^T L)_{m,n} = \sum_j L_{j,m} L_{j,n} = \sum_j p_{j-m} p_{j-n} \tag{5.32}$$

$$= -p_{m-n} + \sum_{j>m} p_{j-m} p_{j-n} = -p_{m-n} + \sum_{l \geq 1} p_l \, p_{(m-n)+l}$$

$$= -p_{m-n} + \sum_{l=1}^{k-(m-n)} p_l \, p_{(m-n)+l},$$

where we make the substitution $l = j - m$ at the next to last step and use the fact that $p_{(m-n)+l} = 0$ when $(m - n) + l > k$.

Since $L^T L$ is symmetric we get the same result when n and m are interchanged. It is clear that when $|m - n| > k$, $L^T L_{m,n} = 0$. This shows that $A = L^T L$. □

The next lemma gives some properties of the matrix A. Note that we are interested in the case in which $\mathcal{V}$ is the potential of a Markov chain. For this to be the case the off diagonal elements on A must be negative.

Lemma 5.3 *Let A be as given in Lemma 5.2 and assume that $\sum_{i=1}^{k} p_i \leq 1$. Then*

$$\sum_{n \in \overline{\mathbb{N}}} A_{m,n} = \left(1 - \sum_{i=1}^{k} p_i \right)^2, \qquad m > k. \tag{5.33}$$

Furthermore, when $p_i \downarrow$,

$$A_{m,n} \leq 0, \qquad \forall n, m \in \overline{\mathbb{N}}, \ n \neq m, \tag{5.34}$$

and

$$\sum_{n \in \overline{\mathbb{N}}} A_{m,n} > \left(1 - \sum_{i=1}^{k} p_i \right)^2, \qquad 1 \leq m \leq k. \tag{5.35}$$

Therefore, $-A$ is a Q-matrix with uniformly bounded entries.

Proof To prove (5.33) we note that by Lemma 5.2, for $m > k$,

$$\sum_{n \in \overline{\mathbb{N}}} A_{m,n} = a_0 + 2 \sum_{j=1}^{k} a_j = 1 + \sum_{i=1}^{k} p_i^2 + 2 \sum_{i=1}^{k} \left(-p_i + \sum_{l=1}^{k-i} p_l \, p_{i+l} \right)$$

$$= \left(1 - \sum_{i=1}^{k} p_i \right)^2 . \tag{5.36}$$

For (5.34) we use (5.11) to see that for all $1 \leq |m - n| \leq k$,

$$a_{|m-n|} = A_{m,n} \leq -p_{|m-n|} + p_{|m-n|+1} \sum_{l=1}^{k-|m-n|} p_l \leq -p_{|m-n|} + p_{|m-n|+1} \leq 0. \tag{5.37}$$

To get (5.35) we note that by (5.17) the row sums of the first k rows of A omit some of the terms a_i, $1 \leq i \leq k$, which are less than or equal to 0.

The final statement in the lemma follows from (5.34), (5.35) and (5.10). □

Remark 5.1 It is clear that $-A$ can be a Q-matrix with uniformly bounded entries, even when $\{p_i\}$ are not decreasing. We see from Example 5.1 that when $k = 2$, $-A$ is always a Q-matrix with uniformly bounded entries. Nevertheless, to keep the statement of Theorem 1.9 from being too cumbersome, we include the hypothesis that $p_i \downarrow$.

The next theorem ties certain k-th order linear regressions to Markov chains.

Theorem 5.1 *Assume that $p_i \downarrow$. Then V is the potential of a Markov chain on $\overline{\mathbb{N}}$ with Q-matrix, $-A$.*

The proof of this theorem depends on the following general result.

Lemma 5.4 *Let Q be the Q-matrix of a transient Markov chain X on $\overline{\mathbb{N}}$ and assume that Q is a $(2m + 1)$-diagonal matrix, with*

$$\sup_{j \in \overline{\mathbb{N}}} |Q_{j,j}| < \infty. \tag{5.38}$$

Let V be a matrix satisfying,

$$V Q = -I, \quad and \quad \sup_{i \in \overline{\mathbb{N}}} |V_{k,i}| < \infty, \quad \forall k \in \overline{\mathbb{N}}. \tag{5.39}$$

Then V is the potential of X, and in particular has positive entries.

If $\sum_i |V_{k,i}| < \infty$, $\forall k \in \overline{\mathbb{N}}$, then the same results hold without the requirement that Q is a $(2m + 1)$-diagonal matrix.

Proof Let U be the potential of X. By (8.1),

$$- \delta_{i,l} = \sum_j Q_{i,j} U_{j,l}. \tag{5.40}$$

Therefore,

$$- V_{k,l} = \sum_i V_{k,i} \sum_j Q_{i,j} U_{j,l} = \sum_i \sum_j V_{k,i} Q_{i,j} U_{j,l}. \tag{5.41}$$

We show immediately below that we can interchange the order of summation. Consequently, by (5.39), for all $k, l \in \overline{\mathbb{N}}$,

$$- V_{k,l} = \sum_j \sum_i V_{k,i} Q_{i,j} U_{j,l} = - \sum_j \delta_{k,j} U_{j,l} = -U_{k,l}. \tag{5.42}$$

This shows that V is the potential of X.

To be able to interchange the order of summation in (5.41), we only need to show that for each fixed k and l,

$$\sum_i \sum_j |V_{k,i}| |Q_{i,j}| |U_{j,l}| < \infty. \tag{5.43}$$

We have $U_{j,l} \le U_{l,l}$ for all j, and for each j there are at most $2m + 1$ elements $|Q_{i,j}|$ that are not equal to 0. Therefore,

$$\sum_i \sum_j |V_{k,i}| |Q_{i,j}| |U_{j,l}| \le U_{l,l} \sum_i \sum_j |V_{k,i}| |Q_{i,j}| = U_{l,l} \sum_j \sum_i |V_{k,i}| |Q_{i,j}|$$

$$\le (2m + 1) U_{l,l} \sup_i |V_{k,i}| \sum_j |Q_{i,j}|. \tag{5.44}$$

Finally, using (1.28) we have

$$\sup_i |V_{k,i}| \sum_j |Q_{i,j}| \le \left(\sup_i |V_{k,i}| \right) 2 \sup_j |Q_{j,j}| < \infty. \tag{5.45}$$

Thus we get (5.43).

If $\sum_i |V_{k,i}| < \infty$, $\forall k \in \overline{\mathbb{N}}$, then in place of (5.44) we have

$$\sum_i \sum_j |V_{k,i}| |Q_{i,j}| |U_{j,l}| \le U_{l,l} \sum_i \sum_j |V_{k,i}| |Q_{i,j}| = U_{l,l} \sum_i |V_{k,i}| \sum_j |Q_{i,j}|$$

$$\le U_{l,l} \left(2 \sup_j |Q_{j,j}| \right) \sum_i |V_{k,i}| < \infty. \tag{5.46}$$

$\square$

Proof of Theorem 5.1 The proof follows immediately from Lemma 5.4 once we show that the hypotheses of the lemma are satisfied. The fact that $-A$ is a Q-matrix is given in Lemma 5.3.

The property that Q is a $(2k + 1)$-diagonal matrix, the condition in (5.38) and the first condition in (5.39) are given in Lemma 5.2.

The second condition in (5.39) is given in (5.7).   □

We now turn to the proof of Theorem 1.9. In this case we need sharp estimates of the covariance $\mathcal{V}$. To this end we introduce a generating function for $\{\phi_n\}$. Set

$$g(x) = \sum_{n=0}^{\infty} \phi_n x^n = \sum_{n=1}^{\infty} \phi_n x^n, \tag{5.47}$$

since $\phi_0 = 0$. It follows from (5.4) that this converges for all $|x| < 1$.

Lemma 5.5 *Let*

$$P(x) = 1 - \sum_{l=1}^{k} p_l x^l. \tag{5.48}$$

Then for all $|x| < 1$,

$$g(x) = \frac{x}{P(x)}. \tag{5.49}$$

Proof We have

$$\sum_{n=1}^{\infty} \phi_n x^n = x + \sum_{n=2}^{\infty} \phi_n x^n = x + \sum_{n=2}^{\infty} \sum_{l=1}^{k} p_l \phi_{n-l} \, x^n \tag{5.50}$$

$$= x + \sum_{l=1}^{k} p_l x^l \sum_{n=2}^{\infty} \phi_{n-l} \, x^{n-l},$$

where we use the fact that $\phi_n = 0$ for $n < 1$. In addition

$$\sum_{n=2}^{\infty} \phi_{n-l} \, x^{n-l} = \sum_{n=2}^{l} \phi_{n-l} \, x^{n-l} + \sum_{n=l+1}^{\infty} \phi_{n-l} \, x^{n-l} \tag{5.51}$$

$$= \sum_{n=l+1}^{\infty} \phi_{n-l} \, x^{n-l} = \sum_{k=1}^{\infty} \phi_k \, x^k = g(x).$$

It follows from (5.50) and (5.51) that,

$$g(x) = x + g(x) \sum_{l=1}^{k} p_l x^l, \tag{5.52}$$

which gives (5.49).   □

Lemma 5.6 *Let $q_1, \ldots, q_k$ be the roots of $P(x)$ which may be complex. Then,*

(i) $\sum_{l=1}^{k} p_l = 1 \iff q_1 = 1$ is a simple root and $|q_l| > 1, l = 2, \ldots, k$.

(ii) $\sum_{l=1}^{k} p_l < 1 \iff |q_l| > 1, \quad l = 1, \ldots, k$.

Proof Assume first that $\sum_{l=1}^{k} p_l = 1$. Then it is obvious that $q_1 = 1$ is a root. Furthermore since,

$$P'(1) = -\sum_{l=1}^{k} l p_l < 0, \tag{5.53}$$

it is not a multiple root. Also, note that

$$1 = \left| \sum_{l=1}^{k} p_l x^l \right| \leq \sum_{l=1}^{k} p_l |x|^l, \tag{5.54}$$

with strict inequality when $|x| = 1$ and $x \neq 1$. Therefore, $|q_l| > 1$ for all $l = 2, \ldots, k$.

If $\sum_{l=1}^{k} p_l < 1$ it is clear from (5.54) that $|q_l| > 1$ for all $l = 1, \ldots, k$. $\square$

We now give a formula for $\phi = \{\phi_n\}$. Define

$$c_1 = \frac{1}{\sum_{l=1}^{k} l p_l}. \tag{5.55}$$

Lemma 5.7 Let $P(x)$ be as given in (5.48) and assume that it has distinct roots q_l of degree $d_l, l = 1, \ldots, k'$.

(i) If $\sum_{l=1}^{k} p_l < 1$, where $k = \sum_{l=1}^{k'} d_l$, then all $|q_l| > 1$ and,

$$\phi_n = \sum_{l=1}^{k'} \sum_{j=1}^{d_l} B_j(q_l) \times \binom{j-1+n}{j-1} \left(\frac{1}{q_l}\right)^n, \tag{5.56}$$

where

$$B_j(q_l) = \frac{(-1)^j}{q_l^j (d_l - j)!} \lim_{x \to q_l} D^{(d_l - j)} \frac{x(x - q_l)^{d_l}}{P(x)}. \tag{5.57}$$

Furthermore,

$$\|\phi\|_1 = \frac{1}{P(1)} < \infty. \tag{5.58}$$

(ii) If $\sum_{l=1}^{k} p_l = 1$ the roots of $P(x)$ can be arranged so that $q_1 = d_1 = 1$ and $|q_l| > 1$, for $l = 2, \ldots, k'$. In this case,

$$\phi_n = c_1 + \psi_n, \tag{5.59}$$

where,

$$\psi_n = \sum_{l=2}^{k'} \sum_{j=1}^{d_l} B_j(q_l) \times \binom{j-1+n}{j-1} \left(\frac{1}{q_l}\right)^n. \tag{5.60}$$

Furthermore,

$$\|\psi\|_1 < \infty. \tag{5.61}$$

Proof Suppose more generally that $P(x)$ is a polynomial with $P(0) \neq 0$ and distinct roots q_l of degree d_l, $l = 1, \ldots, k'$. Then we can write,

$$\frac{x}{P(x)} = \sum_{l=1}^{k'} \sum_{j=1}^{d_l} \frac{a_{l,j}}{(x - q_l)^j} \tag{5.62}$$

where

$$a_{l,j} = \frac{1}{(d_l - j)!} \lim_{x \to q_l} D^{(d_l - j)} \frac{x(x - q_l)^{d_l}}{P(x)}. \tag{5.63}$$

For lack of a suitable reference we provide a simple proof. Let

$$f(x) = \frac{x}{P(x)} - \sum_{l=1}^{k'} \sum_{j=1}^{d_l} \frac{a_{l,j}}{(x - q_l)^j}. \tag{5.64}$$

The function $f(x)$ is a rational function which can only have finite poles at q_l of degrees $\leq d_l$, $l = 1, \ldots, k'$. Consider

$$(x - q_l)^{d_l} f(x) \tag{5.65}$$

$$= \frac{x(x - q_l)^{d_l}}{P(x)} - \sum_{l'=1, l' \neq l}^{k'} \sum_{j=1}^{d_{l'}} \frac{a_{l',j}(x - q_l)^{d_l}}{(x - q_{l'})^j} + \sum_{j=1}^{d_l} a_{l,j}(x - q_l)^{d_l - j}.$$

Considering the definition of the $a_{l,j}$ in (5.63), we see that,

$$\lim_{x \to q_l} D^{(d_l - j)} (x - q_l)^{d_l} f(x) = 0, \tag{5.66}$$

for all $1 \leq j \leq d_l$, and all $l = 1, \ldots, k'$. This shows that the rational function $f(x)$ has no finite poles, which implies that $f(x)$ is a polynomial, and since $\lim_{x \to \infty} f(x) = 0$, we must have $f(x) \equiv 0$. Using (5.64) we get (5.62). Let

$$B_j(q_l) = \frac{a_{l,j}(-1)^j}{q_l^j}. \tag{5.67}$$

Then if all the $|q_l| > 1$ it follows from (5.62) that for all $|x| \leq 1$,

$$\frac{x}{P(x)} = \sum_{l=1}^{k'} \sum_{j=1}^{d_l} \frac{a_{l,j}(-1)^j}{q_l^j (1 - x/q_l)^j} \tag{5.68}$$

$$= \sum_{n=0}^{\infty} \sum_{l=1}^{k'} \sum_{j=1}^{d_l} B_j(q_l) \times \binom{j-1+n}{j-1} \left(\frac{1}{q_l}\right)^n x^n.$$

Therefore, using (5.47) and (5.49) we see that for all $|x| < 1$

$$\sum_{n=0}^{\infty} \phi_n x^n = \sum_{n=0}^{\infty} \sum_{l=1}^{k'} \sum_{j=1}^{d_l} B_j(q_l) \times \binom{j-1+n}{j-1} \left(\frac{1}{q_l}\right)^n x^n. \tag{5.69}$$

This proves (5.56). Since all $|q_l| > 1$ it is clear that (5.69) converges for $x = 1$ so that by combining the last two displays we see that,

$$\frac{1}{P(1)} = \sum_{n=0}^{\infty} \phi_n. \tag{5.70}$$

Since by (5.3), $\phi_n \geq 0$ for all $n \in \overline{\mathbb{N}}$, we get (5.58).

For (ii) we see that as in (5.68) for all $|x| < 1$,

$$\frac{x}{P(x)} = \sum_{n=0}^{\infty} \left(B_1(1) + \sum_{l=2}^{k'} \sum_{j=1}^{d_l} B_j(q_l) \times \binom{j-1+n}{j-1} \left(\frac{1}{q_l}\right)^n \right) x^n, \tag{5.71}$$

where

$$B_1(1) = -\lim_{x \to 1} \frac{x(x-1)}{P(x)} = -\frac{1}{P'(1)} = c_1, \tag{5.72}$$

by L'Hospital's Rule and (5.53). This gives (5.59), in which

$$\psi_n = \sum_{l=2}^{k'} \sum_{j=1}^{d_l} B_j(q_l) \times \binom{j-1+n}{j-1} \left(\frac{1}{q_l}\right)^n. \tag{5.73}$$

Since $|q_l| > 1$, $2 \leq l \leq k'$, it is clear that $\|\psi\|_1 < \infty$. □

Example 5.2 Suppose that $P(x)$ has real roots, a and $-b$, where a has multiplicity 1 and $-b$ has multiplicity 2, and $a \geq 1$. In this case

$$P(x) = -\frac{1}{ab^2}(x-a)(x+b)^2 \tag{5.74}$$

$$= 1 - \frac{1}{ab^2}\left(x^3 + (2b-a)x^2 + (b^2 - 2ab)x\right) \tag{5.75}$$

Therefore,

$$p_1 = \frac{b^2 - 2ab}{ab^2}, \qquad p_2 = \frac{2b - a}{ab^2}, \qquad p_3 = \frac{1}{ab^2}. \tag{5.76}$$

When $b > 2(a + 1)$, $p_1 > p_2 > p_3$. (We know from Lemma 5.6 that we must have $a \geq 1$ and that $\sum_{j=1}^{3} p_j \leq 1$ and is equal to 1 if and only if $a = 1$.)
We have

$$B_1(a) = \frac{(-1)}{a} \lim_{x \to a} \frac{x(-ab^2)}{(x + b)^2} = \frac{ab^2}{(a + b)^2}. \tag{5.77}$$

$$B_1(-b) = \frac{1}{b}(-ab^2) \lim_{x \to -b} D^{(1)} \frac{x}{(x - a)} = \frac{a^2 b}{(a + b)^2} \tag{5.78}$$

$$B_2(-b) = \frac{-ab^2}{b^2} \lim_{x \to -b} \frac{x}{(x - a)} = -\frac{ab}{(a + b)} \tag{5.79}$$

Therefore,

$$\phi_n = \frac{ab^2}{(a + b)^2} \left(\frac{1}{a}\right)^n - \left(\frac{ab}{(a + b)}(n + 1) - \frac{a^2 b}{(a + b)^2}\right) \left(\frac{1}{-b}\right)^n \tag{5.80}$$

$$= \frac{ab^2}{(a + b)^2} \left(\frac{1}{a}\right)^n + (-1)^{n+1} \left(\frac{abn}{(a + b)} + \frac{ab^2}{(a + b)^2}\right) \left(\frac{1}{b}\right)^n \tag{5.81}$$

When $a = 1$ this is,

$$\phi_n = \frac{b^2}{(1 + b)^2} + (-1)^{n+1} \left(\frac{bn}{(1 + b)} + \frac{b^2}{(1 + b)^2}\right) \left(\frac{1}{b}\right)^n. \tag{5.82}$$

One can check that in this case,

$$\sum_{j=1}^{3} j p_j = \frac{(1 + b)^2}{b^2}. \tag{5.83}$$

Proof of Theorem 1.9, (1.64) We use Theorem 1.2. To begin we obtain the denominator in (1.17). Let $\xi = \{\xi_n, n \in \overline{\mathbb{N}}\}$ be a Gaussian sequence defined exactly as $\widetilde{\xi}$ is defined in (5.1) but with the additional conditions that $\sum_{l=1}^{k} p_l = 1$. We now show that

$$\limsup_{n \to \infty} \frac{\xi_n}{(2n \log \log n)^{1/2}} = \frac{1}{\sum_{l=1}^{k} l p_l} \qquad a.s. \tag{5.84}$$

It follows from Lemma 5.7 (ii) that,

$$\phi_n = c_1 + \psi_n, \qquad \text{where} \qquad \psi \in \ell_1^+. \tag{5.85}$$

By (5.5) we can write

$$\xi_n = c_1 S_n + \rho_n, \tag{5.86}$$

where

$$S_n = \sum_{j=1}^{n} g_j \quad \text{and} \quad \rho_n = \sum_{j=1}^{n} \psi_{n+1-j} g_j. \tag{5.87}$$

Note that $E\left(\rho_n^2\right) \le \|\psi\|_2^2$ for all $n \in \overline{\mathbb{N}}$. It follows from the Borel-Cantelli Lemma that,

$$\overline{\lim_{j \to \infty}} \frac{|\rho_j|}{\sqrt{2 \log j}} \le \|\psi\|_2 \quad a.s. \tag{5.88}$$

It now follows from (5.86) and the standard law of the iterated logarithm for S_n that (5.84) holds.

We now show that (1.16) holds. Let $\mathcal{V} = \{V_{j,k}; \ j, k \in \overline{\mathbb{N}}\}$ be as in (5.6). We now find an estimate for the row sums of $(\mathcal{V}(l, n))^{-1}$. For $n \ge k$ set

$$\Xi(l, n) = (\xi_{l+1}, \xi_{l+2}, \dots, \xi_{l+n}) \tag{5.89}$$

and

$$G(l, n) = (\eta_{l+1}, \dots, \eta_{l+k}, g_{l+k+1}, g_{l+k+1}, \dots, g_{l+n}), \tag{5.90}$$

where

$$\eta_{l+j} = \xi_{l+j} - \sum_{i=1}^{j-1} p_i \xi_{l+j-i}, \quad j = 1 \dots, k. \tag{5.91}$$

Note that this is similar in form to (5.1), but starting from $l + 1$.

We use several lemmas. The first one is easy to verify.

Lemma 5.8

$$G(l, n)^T = L(l, n) \Xi(l, n)^T, \tag{5.92}$$

where L is given in (5.19).

It follows from (5.92) that,

$$G(l, n)^T G(l, n) = L(l, n) \Xi(l, n)^T \Xi(l, n) L(l, n)^T. \tag{5.93}$$

We take the expectation of each side and get the vector equation,

$$B \otimes I_{n-k} = L(l, n) W(l, n) L(l, n)^T, \tag{5.94}$$

where

$$B = \text{Cov}(\eta_{l+1}, \dots, \eta_{l+k}).$$

Lemma 5.9

$$(\mathcal{V}(l, n))^{-1} \mathbf{1}_n = \begin{pmatrix} (\mathcal{V}(l, k))^{-1} \mathbf{1}_k \\ 0 \end{pmatrix}, \tag{5.95}$$

where $\mathbf{1}_m$ denotes an m dimensional column vector with all of its components equal to 1.

Note that $(\mathcal{V}(l, n))^{-1} \mathbf{1}_n$ is an n dimensional vector with components that are the row sums of $(\mathcal{V}(l, n))^{-1}$. Therefore, (5.95) states that the first k row sums of $(\mathcal{V}(l, n))^{-1}$ are equal to the row sums of $(\mathcal{V}(l, k))^{-1}$, and the remaining row sums are equal to 0.

Proof Using (5.94) we see that

$$(\mathcal{V}(l, n))^{-1} = L(l, n)^T \left(B^{-1} \otimes I_{n-k} \right) L(l, n). \tag{5.96}$$

In addition, since $L(l, n)$ is a lower triangular matrix we can write it in the block form,

$$L(l, n) = \begin{pmatrix} F & 0 \\ G & H \end{pmatrix}, \tag{5.97}$$

where F is a $k \times k$ matrix. It is easy to check that

$$(L(l, n))^{-1} = \begin{pmatrix} F^{-1} & 0 \\ -H^{-1}GF^{-1} & H^{-1} \end{pmatrix}. \tag{5.98}$$

We also note that since all row sums of $L(l, n)$ after the k-th row are equal to zero,

$$L(l, n)\mathbf{1}_n = \begin{pmatrix} F\mathbf{1}_k \\ 0 \end{pmatrix}. \tag{5.99}$$

It follows from (5.96) that

$$(\mathcal{V}(l, n))^{-1} \mathbf{1}_n = L(l, n)^T \left(B^{-1} \otimes I_{n-k} \right) L(l, n)\mathbf{1}_n. \tag{5.100}$$

Using (5.99) we see that,

$$\left(B^{-1} \otimes I_{n-k} \right) L(l, n)\mathbf{1}_n = \begin{pmatrix} B^{-1}F\mathbf{1}_k \\ 0 \end{pmatrix}. \tag{5.101}$$

Consequently,

$$(\mathcal{V}(l, n))^{-1} \mathbf{1}_n = \begin{pmatrix} F^T & G^T \\ 0 & H^T \end{pmatrix} \begin{pmatrix} B^{-1}F\mathbf{1}_k \\ 0 \end{pmatrix} = \begin{pmatrix} F^T B^{-1}F\mathbf{1}_k \\ 0 \end{pmatrix}. \tag{5.102}$$

On the other hand, by (5.94) ,

$$\begin{pmatrix} F^{-1} & 0 \\ -H^{-1}GF^{-1} & H^{-1} \end{pmatrix} \begin{pmatrix} B & 0 \\ 0 & I \end{pmatrix} \begin{pmatrix} F^{-1} & 0 \\ -H^{-1}GF^{-1} & H^{-1} \end{pmatrix}^T = \mathcal{V}(l, n),$$

from which we obtain

$$F^{-1}B(F^T)^{-1} = \mathcal{V}(l, k). \tag{5.103}$$

Consequently,

$$F^T B^{-1} F = (\mathcal{V}(l, k))^{-1}. \tag{5.104}$$

Using this and (5.102) we get (5.95).   □

We now consider $\mathcal{V}(l, k)$.

Lemma 5.10 *When* $\sum_{l=1}^{k} p_l = 1$,

$$E(\xi_m \xi_n) = c_1^2 (m \wedge n) + a_{n,m}, \tag{5.105}$$

where $|a_{m,n}| \leq D < \infty$, *for all* $m, n \in \overline{\mathbb{N}}$.

Proof By (5.6) and Lemma 5.7 (ii), when $m \leq n$, we have

$$E(\xi_m \xi_n) = \sum_{j=0}^{m-1} \phi_{m-j} \phi_{n-j} \tag{5.106}$$

$$= c_1^2 m + c_1 \sum_{j=0}^{m-1} \psi_{m-j} + c_1 \sum_{j=0}^{m-1} \psi_{n-j}$$

$$+ c_1^2 \sum_{j=0}^{m-1} \psi_{m-j} \psi_{n-j}.$$

Clearly, for all $p \geq m$,

$$\left| \sum_{j=0}^{m-1} \psi_{p-j} \right| \leq \sum_{j=1}^{\infty} |\psi_j| = \|\psi\|_1, \tag{5.107}$$

and,

$$\left| \sum_{j=0}^{m-1} \psi_{m-j} \psi_{n-j} \right| \leq \sum_{j=1}^{\infty} |\psi_j|^2 = \|\psi\|_2^2,$$

where we use the Schwartz Inequality. Combining all these inequalities we see that for $m \leq n$,

$$E(\xi_m \xi_n) = c_1^2 (m \wedge n) + a_{n,m}, \tag{5.108}$$

where,

$$|a_{m,n}| \le 2c_1\|\psi\|_1 + (c_1\|\psi\|_2)^2 := D < \infty. \tag{5.109}$$

□

The next lemma is used to obtain (1.16).

Lemma 5.11 *For all* $1 \le i \le k$,

$$\sum_{j=1}^{k} \mathcal{V}(l,k)^{i,j} = O(1/l). \tag{5.110}$$

Proof It follow from Theorem 5.1 that $\mathcal{V}$ is the potential of a Markov chain . Therefore so is $\mathcal{V}(l,k)$. Consequently, $\mathcal{V}(l,k)^{-1}$ is an M-matrix with positive row sums. This gives the first inequality in (5.111) below,

$$|\mathcal{V}(l,k)^{j,i}| \le \mathcal{V}(l,k)^{j,j} \le A_{j,j} \le 2. \tag{5.111}$$

The second inequality in (5.111) follows from Lemma 5.12, below. The third inequality in (5.111) is given in (5.31).

Clearly,

$$\sum_{j=1}^{k} \mathcal{V}(l,k)_{i,j}\mathcal{V}(l,k)^{j,i} = 1, \qquad 1 \le i \le k. \tag{5.112}$$

Furthermore, by Lemma 5.10,

$$1 = \sum_{j=1}^{k} \mathcal{V}(l,k)_{i,j}\mathcal{V}(l,k)^{j,i} \tag{5.113}$$

$$= c_1^2 \sum_{j=1}^{k}((l+i) \wedge (l+j))\mathcal{V}(l,k)^{i,j} + \sum_{j=1}^{k} a_{l+i,l+j}\mathcal{V}(l,k)^{j,i}$$

$$= c_1^2 l \sum_{j=1}^{k} \mathcal{V}(l,k)^{i,j} + c_1^2 \sum_{j=1}^{k}(i \wedge j)\mathcal{V}(l,k)^{i,j} + \sum_{j=1}^{k} a_{l+i,l+j}\mathcal{V}(l,k)^{j,i}.$$

Therefore,

$$c_1^2 l \sum_{j=1}^{k} \mathcal{V}(l,k)^{i,j} \le 1 + (c_1^2 k + D) \sum_{j=1}^{k} |\mathcal{V}(l,k)^{i,j}| \tag{5.114}$$

$$\le 1 + 2k(c_1^2 k + D),$$

where we use (5.111). This gives (5.110).   □

Lemma 5.12 *Let $X = (\Omega, \mathcal{F}_t, X_t, \theta_t, P^x)$ be a transient symmetric Borel right process with state space $\overline{\mathbb{N}}$, and potential $U = \{U_{j,k}, j, k \in \overline{\mathbb{N}}\}$ and Q-matrix, Q. Assume that*

$$U_{j,k} > 0 \quad and \quad |Q_{j,j}| < \infty, \quad \forall j, k \in \overline{\mathbb{N}}. \tag{5.115}$$

Then for any distinct sequence $l_1, l_2, \ldots, l_n$ in $\overline{\mathbb{N}}$, the matrix $K = \{U_{l_i,l_k}\}_{i,j=1}^n$ is invertible and,

$$K^{j,j} \leq |Q_{l_j,l_j}|, \quad \forall 1 \leq j \leq n. \tag{5.116}$$

Proof We follow the proof of [13, Lemma A.1]. For all $k \in \overline{\mathbb{N}}$ set,

$$L_t^k = \int_0^t 1_{\{X_s = k\}} \, ds. \tag{5.117}$$

It follows from this that for all or all $j, k \in \overline{\mathbb{N}}$ we have,

$$U_{j,k} = E^j \left(L_\infty^k \right). \tag{5.118}$$

Define the stopping time,

$$\sigma = \inf\{t \geq 0 \mid X_t \in \{l_1, l_2, \ldots, l_n, \Delta\} \cap \{X_0\}^c\} \tag{5.119}$$

which may be infinite. By [13, (A.5)],

$$K^{j,j} \leq \frac{1}{E^{l_j} \left(L_\sigma^{l_j} \right)}. \tag{5.120}$$

On the other hand, the amount of time X_t, starting at l_j, remains at l_j is,

$$\sigma_j := \inf\{t \geq 0 \mid X_t \in \{l_j\}^c\}, \tag{5.121}$$

which implies, by (5.117) that,

$$L_{\sigma_j}^{l_j} = \sigma_j. \tag{5.122}$$

In addition, $\sigma_j \leq \sigma$, so that $E^{l_j} \left(L_{\sigma_j}^{l_j} \right) \leq E^{l_j} \left(L_\sigma^{l_j} \right)$. Therefore, it follows from (5.120) and (5.122) that

$$K^{j,j} \leq \frac{1}{E^{l_j} \left(L_{\sigma_j}^{l_j} \right)} = \frac{1}{E^{l_j} (\sigma_j)}. \tag{5.123}$$

Since σ_j is an exponential random variable with mean $1/|Q_{l_j,l_j}|$; (see [17, Chapter 2.6]), we get (5.116). $\qquad\square$

We now consider the potentials corresponding to $\mathcal{V}$.

Lemma 5.13 *Let $f = \mathcal{V}h$, where $h \in \ell_1^+$. Then*

$$f_j = g(j) + \rho_j, \qquad \forall j \in \overline{\mathbb{N}}, \tag{5.124}$$

where g is an increasing strictly concave function and $\sup_j \rho_j = d\|h\|_1$ for some finite constant d.

Proof We show in (5.105) that,

$$\mathcal{V}_{j,k} = c_1^2(j \wedge k) + a_{j,k}, \tag{5.125}$$

where $|a_{j,k}| \leq d$. Therefore

$$f_j = c_1^2 \sum_{k=1}^{\infty} (j \wedge k) h_k + \sum_{k=1}^{\infty} h_k a_{j,k}. \tag{5.126}$$

The lemma now follows from Theorem 2.2. $\qquad\square$

The next lemma shows that (1.16) holds.

Lemma 5.14 *Let $f = \mathcal{V}h$, where $h \in \ell_1^+$. Then*

$$\sum_{j,p=1}^{n} (V(l,n))^{j,p} f_{p+l} = o_l(1), \qquad \text{uniformly in } n. \tag{5.127}$$

Proof It follows from Lemmas 5.9 and 5.11 that for all l sufficiently large, there exists a constant C such that,

$$\sum_{j,p=1}^{n} (V(l,n))^{j,p} f_{p+l} = \sum_{p=1}^{n} f_{p+l} \sum_{j=1}^{n} (V(l,n))^{p,j} \tag{5.128}$$

$$= \sum_{p=1}^{k} f_{p+l} \sum_{j=1}^{k} (V(l,k))^{p,j} \leq C \frac{f_{l+k}}{l}.$$

By Lemma 5.13, $f(j) = o(j)$ and since k is a fixed number, we get (5.127). $\qquad\square$

Proof of Theorem 1.9 (1.64) continued This follows from Theorem 1.2. Lemma 5.14 shows that (1.16) holds. The limit result in (5.84) identifies the denominator in (1.17), and Lemma 5.13 gives (1.18). $\qquad\square$

Proof of Theorem 1.9, (1.62) This follows from Theorem 1.3. We show that the hypotheses in (1.20) are satisfied. It follows from (5.7) that $\inf_{j\geq 1} \mathcal{V}_{j,j}$

$= 1$. Therefore, the first condition in (1.20) is satisfied. In addition, by (5.6), when $n \geq m$,

$$V_{m,n} = \sum_{j=0}^{m-1} \phi_{m-j}\phi_{n-j} = \sum_{j=1}^{m} \phi_j \phi_{n-m+j} \tag{5.129}$$

Therefore,

$$\sum_{n=m}^{\infty} V_{m,n} = \sum_{j=1}^{m} \phi_j \sum_{n=m}^{\infty} \phi_{n-m+j} \leq \|\phi\|_1^2. \tag{5.130}$$

Obviously, this holds when $n < m$ so we see that the second condition in (1.20) is also satisfied.

Furthermore, we see that

$$\lim_{n\to\infty} V_{n,n} = \|\phi\|_2^2 := c^*. \tag{5.131}$$

Therefore, (1.62) follows from Theorem 1.3.

To obtain the upper bound in (1.63) we note that by (5.1),

$$E\left(\tilde{\xi}_n^2\right) = E\left(\sum_{l=1}^{k} p_l \tilde{\xi}_{n-l}\right)^2 + 1 \tag{5.132}$$

$$= \sum_{l,l'=1}^{k} p_l p_{l'} E(\tilde{\xi}_{n-l}\tilde{\xi}_{n-l'}) + 1$$

$$\leq \left(\sum_{l=1}^{k} p_l\right)^2 E\left(\tilde{\xi}_n^2\right) + 1.$$

Here we use the Cauchy-Schwarz Inequality and the fact that $E(\xi_n^2) \uparrow$ to get,

$$E(\tilde{\xi}_{n-l}\tilde{\xi}_{n-l'}) \leq \left(E(\tilde{\xi}_{n-l}^2)E(\tilde{\xi}_{n-l'}^2)\right)^{1/2} \leq E\left(\tilde{\xi}_n^2\right). \tag{5.133}$$

The lower bound is obtained from (5.3). We can add additional terms in situations where it is useful.

The fact that $f \in c_0^+$ if and only if $f = Vh$, where $h \in c_0^+$ follows from Lemma 7.2 once we show that (7.20) holds. To see this we note that.

$$\sum_{m=1}^{n/2} V_{m,n} = \sum_{m=1}^{n/2}\sum_{j=1}^{m} \phi_j \phi_{n-m+j} = \sum_{j=1}^{m} \phi_j \sum_{m=1}^{n/2} \phi_{n-m+j}$$

$$\leq \|\phi\|_1 \sum_{k=n/2}^{\infty} \phi_k. \tag{5.134}$$

Remark 5.2 It follows from Lemma 5.7 (i) that   □

$$c^* = \|\phi\|_2^2 = \sum_{l,l'=1}^{k'} \sum_{j=1}^{d_l} \sum_{j'=1}^{d_{l'}} B_j(q_l) B_{j'}(q_{l'}) F_{j,j'}(q_l q_{l'}), \tag{5.135}$$

where $B_j(q_l)$ is given in (5.57) and

$$F_{j,j'}(q_l q_{l'}) = \sum_{n=0}^{\infty} \binom{j-1+n}{j-1} \binom{j'-1+n}{j'-1} \left(\frac{1}{q_l q_{l'}}\right)^n. \tag{5.136}$$

Example 5.3 Suppose that

$$P(x) = -\frac{1}{ab}(x-a)(x+b) = 1 - \frac{1}{ab}(x^2 + (b-a)x) \tag{5.137}$$

where $a > 1$ and $b \geq a$. This assures us that p_1 and $p_2 > 0$ and that $p_1 + p_2 < 1$.
   We have

$$B_1(a) = \frac{-1}{a} \lim_{x \to a} \frac{x(x-a)(-ab)}{P(x)} = \frac{ab}{a+b}. \tag{5.138}$$

Similarly,

$$B_1(-b) = \frac{1}{b} \lim_{x \to -b} \frac{x(x+b)(-ab)}{P(x)} = -\frac{ab}{a+b}. \tag{5.139}$$

Consequently,

$$c^* = \left(\frac{ab}{a+b}\right)^2 \left(F_{1,1}(a^2) + F_{1,1}(b^2) - 2F_{1,1}(a(-b))\right), \tag{5.140}$$

$$= \left(\frac{ab}{a+b}\right)^2 \left(\frac{a^2}{a^2-1} + \frac{b^2}{b^2-1} - 2\frac{ab}{ab+1}\right), \tag{5.141}$$

   For a concrete example suppose that $a = -1 + \sqrt{5}$ and $-b = -(1 + \sqrt{5})$. (These are the roots of $x^2/4 + x/2 - 1$.) Then,

$$\|\phi\|_2^2 = \frac{4}{5}\left(\frac{6 - 2\sqrt{5}}{5 - 2\sqrt{5}} + \frac{6 + 2\sqrt{5}}{5 + 2\sqrt{5}} - \frac{8}{5}\right) = \frac{48}{25} \approx 1.92.$$

(The bound in (1.63) is $16/7 \approx 2.28$.)

Proof of Theorem 1.10 Consider $\{\widetilde{\mathcal{Y}}_{\alpha, t_j}, j \in \overline{\mathbb{N}}\}$. This is an α-permanental sequence with kernel,

$$\widetilde{\mathcal{V}}_{t_j, t_k} = \mathcal{V}_{t_j, t_k} + f_{t_k}, \qquad j, k \in \overline{\mathbb{N}}. \tag{5.142}$$

It follows from (5.105) that for an increasing sequence $\{t_j\}$,

$$\tilde{\mathcal{V}}_{t_j,t_k} = c_1^2(t_j \wedge t_k) + O(1) + f_{t_k}, \qquad j,k \in \overline{\mathbb{N}}. \tag{5.143}$$

Set

$$\widehat{\mathcal{V}}_{t_j,t_k} = \frac{\tilde{\mathcal{V}}_{t_j,t_k}}{(\tilde{\mathcal{V}}_{t_j,t_j})^{1/2}(\tilde{\mathcal{V}}_{t_k,t_k})^{1/2}}. \tag{5.144}$$

For $t_j \leq t_k$ we have,

$$\widehat{\mathcal{V}}_{t_j,t_k} + \widehat{\mathcal{V}}_{t_k,t_j} = \frac{2t_j + O(1) + f_{t_k} + f_{t_j}}{(t_j t_k)^{1/2}}. \tag{5.145}$$

Using the hypothesis that $f_j = o(j^{1/2})$ we see that for $t_j \leq t_k$,

$$\widehat{\mathcal{V}}_{t_j,t_k} + \widehat{\mathcal{V}}_{t_k,t_j} = 2\left(\frac{t_j}{t_k}\right)^{1/2} + o(1), \qquad as\, t_j \to \infty. \tag{5.146}$$

In particular if $t_j = \theta^j$ for some $\theta > 1$, for all $j \in \overline{\mathbb{N}}$, we have

$$\widehat{\mathcal{V}}_{t_j,t_k} + \widehat{\mathcal{V}}_{t_k,t_j} = 2\theta^{-|k-j|/2} + o(1), \qquad as\, j,k \to \infty., \tag{5.147}$$

Also, it is easy to see that,

$$\widehat{\mathcal{V}}_{\theta^j,\theta^k} - \widehat{\mathcal{V}}_{\theta^k,\theta^j} = o(1), \qquad as\, j,k \to \infty. \tag{5.148}$$

The estimates in (5.147) and (5.148) enable us to show that the hypotheses in [15, Lemma 7.1] are satisfied. Therefore, by taking θ sufficiently large we have that any $\epsilon > 0$,

$$\limsup_{j\to\infty} \frac{\tilde{\mathcal{Y}}_{\alpha,\theta^j}}{\theta^j \log j} \geq 1 - \epsilon. \tag{5.149}$$

This gives the lower bound in (1.64) for all $\alpha > 0$.   □

Extending the generalization of first order linear regressions in (4.72), we generalize the class of higher order Gaussian autoregressive sequences and find their covariances. In the beginning of this section we consider a class of k−th order autoregressive Gaussian sequences, $\overline{\xi} = \{\overline{\xi}_n, n \in \overline{\mathbb{N}}\}$, for $k \geq 2$. Let $g_1, g_2, \ldots$ be independent standard normal random variables and let $p_i > 0$, $i = 1, \ldots, k$, with $\sum_{l=1}^k p_l \leq 1$. We define the Gaussian sequence $\overline{\xi} = \{\overline{\xi}_n, n \in \overline{\mathbb{N}}\}$ by,

$$\overline{\xi}_1 = \frac{g_1}{a}, \quad \text{and} \quad \overline{\xi}_n = \sum_{l=1}^k p_l \overline{\xi}_{n-l} + g_n, \qquad n \geq 2, \tag{5.150}$$

where $\overline{\xi}_i = 0$ for all $i \leq 0$ and $a \neq 0$.

Lemma 5.15

$$\mathcal{V}_{m,n}^{[a^2]} := E(\bar{\xi}_m \bar{\xi}_n) = \frac{1 - a^2}{a^2} \phi_m \phi_n + E\left(\widetilde{\xi}_m \widetilde{\xi}_n\right) \tag{5.151}$$

$$= \frac{1 - a^2}{a^2} \mathcal{V}_{m,1} \mathcal{V}_{1,n} + \mathcal{V}_{m,n}.$$

Furthermore, for all $j \in \overline{\mathbb{N}}$,

$$\lim_{j \to \infty} \frac{\mathcal{V}_{j,j}^{[a^2]}}{\mathcal{V}_{j,j}} = 1. \tag{5.152}$$

Proof Generalizing (5.5) in Lemma 5.1 we have,

$$\bar{\xi}_n = \phi_n \frac{g_1}{a} + \sum_{j=2}^{n} \phi_{n+1-j} g_j, \tag{5.153}$$

where the ϕ_n are defined in (5.2) for $\bar{\xi}_n$, not $\bar{\xi}_n$. The only difference between this and (5.5) is that g_1 is replaced by g_1/a. Therefore, it follows from this and (5.6) that,

$$E(\bar{\xi}_m \bar{\xi}_n) = \frac{\phi_n \phi_m}{a^2} + \sum_{j=2}^{m \wedge n} \phi_{m+1-j} \phi_{n+1-j} \tag{5.154}$$

$$= \frac{1 - a^2}{a^2} \phi_m \phi_n + E\left(\widetilde{\xi}_m \widetilde{\xi}_n\right).$$

The last equation in (5.151) follows from (5.6).

To obtain (5.152) we note that by (5.151),

$$\mathcal{V}_{j,j}^{[a^2]} = \mathcal{V}_{j,j} + \frac{1 - a^2}{a^2} \phi_j^2. \tag{5.155}$$

If $\sum_{j=1}^{k} p_j < 1$ it follows from (5.58) that $\{\phi_j\} \in \ell_1$. This gives (5.152) in this case. When $\sum_{j=1}^{k} p_j = 1$ it follows from (5.59) and (5.60) that $\phi_j = c_1 + \psi_j$ where $\{\psi_j\} \in \ell_1$. Since $\lim_{j \to \infty} \mathcal{V}_{j,j} = \infty$ in this case we also get (5.152). $\qquad \square$

We now show that $\mathcal{V}_{m,n}^{[a^2]}$ is the potential of a transient Markov chain. For the reason given in Remark 5.1 we assume that $p_i \downarrow$.

Consider the matrix A defined in Lemma 5.2. We generalize this matrix by replacing $A_{1,1} = 1 + \sum_{i=1}^{k} p_i^2$ by $a^2 + \sum_{i=1}^{k} p_i^2$. Denote the generalized matrix by $A^{[a^2]}$. In this notation $A = A^{[1]}$.

Theorem 5.2 *If*

$$a^2 \geq \frac{1}{2} \left(\sum_{i=1}^{k} p_i \left(2 - \sum_{i=1}^{k} p_i\right) - \sum_{i=1}^{k} p_i^2 \right), \tag{5.156}$$

then $|a| > 0$ and $-A^{[a^2]}$ is the Q-matrix of a transient Markov chain $\mathcal{Y}^{[a^2]}$ with potential $\{\mathcal{V}^{[a^2]}_{m,n};\ m, n \in \overline{\mathbb{N}}\}$.

Proof We show in Lemma 5.3, (5.35), that the m−th row sums of the $-A^{[a^2]}$, $2 \le m \le k$ are strictly greater than 0. Therefore, to see that $-A^{[a^2]}$ is a Q-matrix of a transient Markov chain, it suffices to check that the first row sum of $A^{[a^2]}$ is greater than or equal to 0. We write this row sum as,

$$\sum_{j=1}^{\infty} A^{[a^2]}_{1,j} = a^2 + \sum_{i=1}^{k} p_i^2 + \gamma, \tag{5.157}$$

where γ is the sum of all terms to the right of the diagonal. It follows from (5.36) that,

$$1 + \sum_{i=1}^{k} p_i^2 + 2\gamma = \left(1 - \sum_{i=1}^{k} p_i\right)^2. \tag{5.158}$$

Therefore,

$$\gamma = -\frac{1}{2}\left(1 + \sum_{i=1}^{k} p_i^2 - \left(1 - \sum_{i=1}^{k} p_i\right)^2\right). \tag{5.159}$$

By (5.157) we see that the first row sum of $A^{[a^2]}$ is strictly greater than zero if,

$$a^2 \ge -\gamma - \sum_{i=1}^{k} p_i^2, \tag{5.160}$$

which gives (5.156).

Note that

$$-\gamma - \sum_{i=1}^{k} p_i^2 = \frac{1}{2}\left(\sum_{i=1}^{k} p_i \left(2 - \sum_{i=1}^{k} p_i\right) - \sum_{i=1}^{k} p_i^2\right). \tag{5.161}$$

It is easy to see that unless $p_1 = 1$ the right-hand side of (5.160) is strictly greater than 0. Since this is not possible by hypothesis, we see that $|a| > 0$. □

Assume that (5.156) holds. As in the proof of Theorem 5.1, to show that $\mathcal{V}^{[a^2]}$ is the potential for the Markov chain with Q-matrix $-A^{[a^2]}$ it suffices to show that

$$\mathcal{V}^{[a^2]} A^{[a^2]} = I. \tag{5.162}$$

Using (5.151) we see that (5.162) can be written as,

$$\sum_{j=1}^{\infty} \left(\frac{1-a^2}{a^2} \mathcal{V}_{m,1} \mathcal{V}_{1,j} + \mathcal{V}_{m,j} \right) \left(A_{j,n} + (a^2 - 1)\delta_1(j)\delta_1(n) \right) = \delta_{m,n}. \quad (5.163)$$

Since $\mathcal{V}A = I$ by (5.14) and $\mathcal{V}_{1,1} = 1$, we need only show that for all m,

$$\frac{1-a^2}{a^2} \mathcal{V}_{m,1} \sum_{j=1}^{\infty} \mathcal{V}_{1,j} A_{j,n} + \frac{1-a^2}{a^2} (a^2 - 1)\mathcal{V}_{m,1}\delta_1(n) + (a^2 - 1)\mathcal{V}_{m,1}\delta_1(n) = 0,$$

which follows easily since $\sum_{j=1}^{\infty} \mathcal{V}_{1,j} A_{j,n} = \delta_1(n)$.   □

We use Theorem 5.2 to extend Theorem 1.9 to potentials of the form $\mathcal{V}^{[a^2]}$.

Theorem 5.3 *Suppose that a^2 satisfies (5.156). Then Theorem 1.9 holds with $\mathcal{Y}$ and $\mathcal{V}$ replaced by $\mathcal{Y}^{[a^2]}$ and $\mathcal{V}^{[a^2]}$.*

Proof The analogue of (1.62) follows from Theorem 1.3 as in the proof of Theorem 1.9, (i). We now verify that the conditions for Theorem 1.3 are satisfied. By (5.151)

$$\mathcal{V}_{i,j}^{[a^2]} = \mathcal{V}_{i,j} + \frac{1-a^2}{a^2} \phi_i \phi_j. \quad (5.164)$$

Therefore, by (5.130) and the fact that $\phi_i \leq 1$ for all $i \in \overline{\mathbb{N}}$,

$$\sum_{j=1}^{\infty} \mathcal{V}_{i,j}^{[a^2]} = \sum_{j=1}^{\infty} \mathcal{V}_{i,j} + \frac{1-a^2}{a^2} \phi_i \sum_{j=1}^{\infty} \phi_j \quad (5.165)$$

$$\leq 2\|\phi\|_1^2 + \frac{1-a^2}{a^2} \|\phi\|_1^2 = \frac{1+a^2}{a^2} \|\phi\|_1^2.$$

Therefore, $\mathcal{V}^{[a^2]}$ satisfies the second condition in (1.20).

Since $-A^{[a^2]}$ is a Q-matrix, $\mathcal{V}_{j,j}^{[a^2]} > 0$ for each $j \in \overline{\mathbb{N}}$. In addition it follows from (5.130) and (5.152) that

$$\lim_{j \to \infty} \mathcal{V}_{j,j}^{[a^2]} = c^*. \quad (5.166)$$

Therefore, $\mathcal{V}^{[a^2]}$ also satisfies the first condition in (1.20). Using (5.166) and Theorem 1.3 we get the analogue (1.62).

The proof of the analogue of (1.64) follows from a slight generalization of the proof of Theorem 1.9, (ii). We find an estimate for the row sums of $\left(\mathcal{V}^{[a^2]}(l, n) \right)^{-1}$. Consider the terms defined in (5.89)–(5.91) but with ξ replaced by $\overline{\xi}$ defined in (5.150). Lemmas 5.8 and 5.9 continue to hold with this substitution. The next lemma gives an analogue of Lemma 5.10.

Lemma 5.16 *When* $\sum_{l=1}^{k} p_l = 1$,

$$E\left(\bar{\xi}_m \bar{\xi}_n\right) = c_1^2(m \wedge n) + a'_{n,m},$$ (5.167)

where $|a'_{m,n}| \leq D' < \infty$, *for all* $m, n \in \overline{\mathbb{N}}$.

Proof This follows immediately from (5.154), (5.105) and then the fact that $\phi_i \leq 1$ for all $i \in \overline{\mathbb{N}}$.   □

Proof of Theorem 5.3 continued Using Lemma 5.16 and following the proof of Lemma 5.11 we see that (5.110) holds for $\mathcal{V}^{[a^2]}$. Similarly, Lemmas 5.13 and 5.14 hold for $\mathcal{V}^{[a^2]}$. Consequently the proof of the analogue (1.64) follows immediately from the proof of Theorem 1.9.   □

Theorem 1.10 also holds for potentials of the form $\mathcal{V}^{[a^2]}$.

Theorem 5.4 *Under the hypotheses of Theorem 5.3 assume in addition that* $f_j = o(j^{1/2})$ *as* $j \to \infty$. *Then the analogue (1.64) holds for all* $\alpha > 0$.

Proof This is immediate since Lemma 5.16 gives (5.143).   □

Remark 5.3 Similar to what we point out in Remark 4.3 the condition in (5.156) is necessary for $\mathcal{V}^{[a^2]}$ to be the potential of a Markov chain whereas (5.151) holds for all $a \neq 0$.

Chapter 6
Relating Permanental Sequences to Gaussian Sequences

Theorem 1.2 is a fundamental result that relates the asymptotic behavior of certain permanental sequences to related Gaussian sequences. In this chapter we show how this is done. Let $H = \{H_{j,k}; j, k = 1, \ldots, n\}$ be an $n \times n$ matrix with positive entries. We define,

$$H_{Sym} = \{(H_{i,j} H_{j,i})^{1/2}\}_{i,j=1}^{n}. \tag{6.1}$$

Let K be an $n \times n$ inverse M-matrix and let $A = K^{-1}$. We define

$$A_{sym} = \begin{cases} A_{j,j} & j = 1, \ldots, n \\ -(A_{i,j} A_{j,i})^{1/2} & i, j = 1, \ldots, n, i \neq j \end{cases}, \tag{6.2}$$

and

$$K_{isymi} = (A_{sym})^{-1}. \tag{6.3}$$

(The notation $isymi$ stands for: take the inverse, symmetrize and take the inverse again.) Obviously, when K is symmetric, $A_{sym} = A$ and $K_{isymi} = K$, but when K is not symmetric, $K_{isymi} \neq K$.

Lemma 6.1 *The matrix K_{isymi} is an inverse M-matrix and, consequently is the kernel of α-permanental random variables.*

Proof The matrix $A = K^{-1}$ is a non-singular M-matrix. Therefore, by [14, Lemma 3.3], A_{sym} is a non-singular M-matrix. We denote its inverse by K_{isymi}. The fact that K_{isymi} is the kernel of α-permanental random variables follows from [5, Lemma 4.2]. □

Theorem 1.2 is an application of the next lemma which is [14, Corollary 3.1].

M. B. Marcus and J. Rosen, *Asymptotic Properties of Permanental Sequences*, SpringerBriefs in Probability and Mathematical Statistics, https://doi.org/10.1007/978-3-030-69485-2_6

Lemma 6.2 *For any $\alpha > 0$ let $\widetilde{X}_\alpha = (\widetilde{X}_{\alpha,0}, \widetilde{X}_{\alpha,1}, \ldots, \widetilde{X}_{\alpha,n})$ be an α-perman-ental random variable with kernel $K(n+1)$ that is an inverse M-matrix and set $A(n+1) = K(n+1)^{-1}$. Let $\widetilde{Y}_\alpha = (\widetilde{Y}_{\alpha,0}, \widetilde{Y}_{\alpha,1}, \ldots, \widetilde{Y}_{\alpha,n})$ be the α-permanental random variable determined by $K(n+1)_{isymi}$. Then for all functions g of $\widetilde{X}_\alpha(n+1)$ and $\widetilde{Y}_\alpha(n+1)$ and sets $\mathcal{B}$ in the range of g,*

$$\frac{|A(n+1)|^\alpha}{|A(n+1)_{sym}|^\alpha} P\left(g(\widetilde{Y}_\alpha(n+1)) \in \mathcal{B}\right) \leq P\left(g(\widetilde{X}_\alpha(n+1)) \in \mathcal{B}\right) \quad (6.4)$$

$$\leq \left(1 - \frac{|A(n+1)|^\alpha}{|A(n+1)_{sym}|^\alpha}\right) + \frac{|A(n+1)|^\alpha}{|A(n+1)_{sym}|^\alpha} P\left(g(\widetilde{Y}_\alpha(n+1)) \in \mathcal{B}\right).$$

It is clear that for this lemma to be useful we would like to have $|A(n+1)|^\alpha/|A(n+1)_{sym}|^\alpha$ close to 1.

To obtain limit theorems we apply this lemma to sequences $\widetilde{X}_\alpha(l, n+1) = (\widetilde{X}_{\alpha,l}, \widetilde{X}_{\alpha,l+1}, \ldots, \widetilde{X}_{\alpha,l+n})$ with kernels $K(l, n+1)$ and consider

$$\nu_{l,n} := \frac{|A(l, n+1)_{sym}|}{|A(l, n+1)|}, \quad (6.5)$$

where $A(l, n+1) = (K(l, n+1))^{-1}$. $(\widetilde{Y}_\alpha(l, n+1)$ is the α-permanental random variable determined by $K(l, n+1)_{isymi}$.)

Here is how we obtain the matrices $K(l, n+1)$. We start with a transient symmetric Borel right process, say X, with state space $\overline{\mathbb{N}}$, and potential $U = \{U_{j,k}\}_{j,k=1}^\infty$. Then by [13, Lemma A.1],

$$U(l, n) = \{U_{l+j,l+k}\}_{j,k=1}^n, \quad (6.6)$$

is the potential of a transient symmetric Borel right process, say $\widehat{X}$ on $\{1, \ldots, n\}$. This implies that $U(l, n)$ is a symmetric inverse M matrix with positive row sums, i.e., $\sum_{k=1}^n (U(l, n))^{j,k} \geq 0$, for all $1 \leq j \leq n$.

Let $f = \{f_n\}_{n \in \overline{\mathbb{N}}}$ be an excessive function with respect to X. It follows from Theorem 1.1 that,

$$\widetilde{U}(l, n) = \{U_{l+j,l+k} + f_{l+k}\}_{j,k=1}^n \quad (6.7)$$

is the kernel of an α-permanental vector. We define $K(l, n+1)$ to be an extension of $\widetilde{U}(l.n)$ in the following way:

$$\begin{aligned}
K(l, n+1)_{j,0} &= 1, & j &= 0, \ldots, n, \\
K(l, n+1)_{0,k} &= f_{l+k}, & k &= 1, \ldots, n, \\
K(l, n+1)_{j,k} &= U_{l+j,l+k} + f_{l+k}, & j, k &= 1, \ldots, n.
\end{aligned} \quad (6.8)$$

Written out this is,

$$K(l, n+1) = \begin{pmatrix} 1 & f_{l+1} & \cdots & f_{l+n} \\ 1 & U_{l+1,1} + f_{l+1} & \cdots & U_{l+1,n} + f_{l+n} \\ \vdots & \vdots & \ddots & \vdots \\ 1 & U_{l+n,1} + f_{l+1} & \cdots & U_{l+n,n} + f_{l+n} \end{pmatrix}. \tag{6.9}$$

It is clear from (6.9), by subtracting the first row from all other rows, that,

$$|K(l, n+1)| = |U(l, n)|. \tag{6.10}$$

Therefore $K(l, n+1)$ is invertible. Let $A(l, n+1) = K(l, n+1)^{-1}$. By multiplying the following matrix on the right by $K(l, n+1)$ one can check that,

$A(l, n+1) =$

$$\begin{pmatrix} 1 + \rho_{l,n} & -\sum_{j=1}^{n}(U(l,n))^{j,1} f_{l+j} \cdots & -\sum_{j=1}^{n}(U(l,n))^{j,n} f_{l+j} \\ -\sum_{k=1}^{n}(U(l,n))^{1,k} & U(l,n)^{1,1} & \cdots & U(l,n)^{1,n} \\ \vdots & \vdots & \ddots & \vdots \\ -\sum_{k=1}^{n}(U(l,n))^{n,k} & U(l,n)^{n,1} & \cdots & U(l,n)^{n,n} \end{pmatrix} \tag{6.11}$$

where

$$\rho_{l,n} = \sum_{j,k=1}^{n} (U(l,n))^{j,k} f_{l+k}. \tag{6.12}$$

Note that all the row sums of $A(l, n+1)$ are equal to 0, except for the first row sum which is equal to 1. Also the terms $U(l, n)^{j,k}$, $j, k = 1, \ldots, n$, $j \neq k$ are negative because $U(l, n)$ is an inverse M matrix. Therefore, to show that $A(l, n+1)$ is an M-matrix with positive row sums we need only check that

$$\sum_{j=1}^{n}(U(l,n))^{j,k} f_{l+j} \geq 0, \qquad \forall 1 \leq k \leq n. \tag{6.13}$$

We first consider the case in which,

$$f = Uh, \qquad h \in l_1^+. \tag{6.14}$$

We point out in the second paragraph after the statement of Theorem 1.1 that in this case $f_j < \infty$, for all $j \in \overline{\mathbb{N}}$.

It follows from [15, Theorem 6.1] applied to the transient symmetric Borel right process Z, with state space $\overline{\mathbb{N}}$ and potential function f in (6.14) that we can obtain a transient symmetric Borel right process $\widetilde{Z}$, with state space $\overline{\mathbb{N}} \cup *$, where $*$ is an isolated point, such that $\widetilde{Z}$ has potential

$$\tilde{U}_{j,k} = U_{j,k} + f_k, \qquad j, k \in \bar{\mathbb{N}} \tag{6.15}$$
$$\tilde{U}_{*,k} = f_k, \quad \text{and} \quad \tilde{U}_{j,*} = \tilde{U}_{*,*} = 1.$$

It then follows from [13, Lemma A.1] that $K(l, n + 1)$, defined in (6.8), is invertible and its inverse, $A(l + n)$ is a nonsingular M matrix, so (6.13) holds. The inequality in (6.13) can be extended to hold for all excessive functions because any excessive function is the increasing limit of potential functions $\{f^{(m)}\}$ of the form (6.14). (See the proof of [15, Theorem 1.11].)

Remark 6.1 The reader may wonder why we work with $K(l, n + 1)$ instead of simply $\{U_{l+j,1} + f_{l+k}\}_{j,k=1}^{n}$. It is because it is easy to find $(K(l, n + 1))^{-1}$ and it turns out to be a simple modification of $U(l, n)^{-1}$. This is not the case for the inverse of $\{U_{l+j,1} + f_{l+k}\}_{j,k=1}^{n}$.

The next lemma is the critical estimate in the proof of Theorem 1.2.

Lemma 6.3 *For the matrices $A(l, n + 1)$ and $A(l, n + 1)_{sym}$,*

$$1 \leq \nu_{l,n} \leq 1 + \rho_{l,n}. \tag{6.16}$$

Proof It follows from (6.10) that

$$|A(l, n + 1)| = |(U(l, n))^{-1}|. \tag{6.17}$$

Also, since U is symmetric,

$$A(l, n + 1)_{sym} = \begin{pmatrix} 1 + \rho_{l,n} & -m(l, n)_1 & \cdots & -m(l, n)_n \\ -m(l, n)_1 & U(l, n)^{1,1} & \cdots & U(l, n)^{1,n} \\ \vdots & \vdots & \ddots & \vdots \\ -m(l, n)_n & U(l, n)^{n,1} & \cdots & U(l, n)^{n,n} \end{pmatrix}, \tag{6.18}$$

where

$$m(l, n)_k = (c(l, n)_k r(l, n)_k)^{1/2}, \tag{6.19}$$

and

$$c(l, n)_k = \sum_{j=1}^{n} (U(l, n))^{j,k} f_{l+j}, \quad \text{and} \quad r(l, n)_k = \sum_{j=1}^{n} (U(l, n))^{k,j}. \tag{6.20}$$

We write this in block form,

$$A(l, n + 1)_{sym} = \begin{pmatrix} (1 + \rho_{l,n}) & -\mathbf{m}(l, n) \\ -\mathbf{m}(l, n)^T & U(l, n)^{-1} \end{pmatrix}, \tag{6.21}$$

where $\mathbf{m}(l, n) = (m(l, n)_1, \ldots, m(l, n)_n)$. Therefore,

$$|A(l, n+1)_{sym}| = |U(l, n)^{-1}| \left((1 + \rho_{l,n}) - \mathbf{m}(l, n)U(l, n)\mathbf{m}(l, n)^T \right). \quad (6.22)$$

(See, e.g., [3, Appendix B].)

Using this and (6.17) we see that

$$v_{l,n} = (1 + \rho_{l,n}) - \mathbf{m}(l, n)U(l, n)\mathbf{m}(l, n)^T. \quad (6.23)$$

It follows from [14, Lemma 3.3] that $v_{l,n} \geq 1$. Furthermore, since $U(l, n)$ is positive, $\mathbf{m}(l, n)U(l, n)\mathbf{m}(l, n)^T \geq 0$. This gives (6.16). $\qquad \square$

The next lemma gives another critical estimate. Recall that $K(l, n+1)_{isymi}$ is defined to be $(A(l, n+1))_{sym}^{-1})^{-1}$. It is an $(n+1) \times (n+1)$ matrix indexed by $j, k = 0, \ldots, n$. In the next lemma we consider the $n \times n$ matrix $\{K_{isymi}(l, n+1)\}_{j,k=1}^n$.

Lemma 6.4

$$\{K_{isymi}(l, n+1)\}_{j,k=1}^n = \{U(l, n)_{j,k} + a(l, n)_j a(l, n)_k\}_{j,k=1}^n \quad (6.24)$$

where

$$a(l, n)_j = v_{l,n}^{-1/2} (\mathbf{m}(l, n)U(l, n))_j \leq f_{l+j}^{1/2}, \quad 1 \leq j \leq n. \quad (6.25)$$

Proof By (6.21) and the formula for the inverse of $A(l, n+1)_{sym}$ written as a block matrix; (see, e.g., [3, Appendix B]), we have,

$$K(l, n+1)_{isymi} \qquad\qquad\qquad\qquad\qquad\qquad\qquad\qquad (6.26)$$
$$= \begin{pmatrix} v_{l,n}^{-1} & v_{l,n}^{-1}\mathbf{m}(l, n)U(l, n) \\ v_{l,n}^{-1}U(l, n)\mathbf{m}(l, n)^T & U(l, n) + v_{l,n}^{-1}U(l, n)\mathbf{m}(l, n)^T\mathbf{m}(l, n)U(l, n) \end{pmatrix}.$$

Note that for $i, j = 1 \ldots, n$,

$$\left(U(l, n) + v_{l,n}^{-1}U(l, n)\mathbf{m}(l, n)^T\mathbf{m}(l, n)U(l, n) \right)_{i,j} = U(l, n)_{i,j} + a(l, n)_i a(l, n)_j. \quad (6.27)$$

Using the fact that $U(l, n) \geq 0$, we see that,

$$(\mathbf{m}(l, n)U(l, n))_j = \sum_{i=1}^n m(l, n)_i U(l, n)_{i,j} \qquad\qquad (6.28)$$

$$= \sum_{i=1}^n (c(l, n)_i r(l, n)_i)^{1/2} \, U(l, n)_{i,j}$$

$$\leq \left(\sum_{i=1}^n c(l, n)_i U(l, n)_{i,j} \right)^{1/2} \left(\sum_{i=1}^n r(l, n)_i U(l, n)_{i,j} \right)^{1/2}.$$

Furthermore,

$$\sum_{i=1}^{n} c(l,n)_i U(l,n)_{i,j} = \sum_{i=1}^{n}\sum_{k=1}^{n} (U(l,n))^{k,i} f_{l+k} U(l,n)_{i,j} \qquad (6.29)$$

$$= \sum_{k=1}^{n} f_{l+k} \sum_{i=1}^{n} (U(l,n))^{k,i} U(l,n)_{i,j}$$

$$= \sum_{k=1}^{n} f_{l+k}\delta_{k,j} = f_{l+j},$$

and, similarly,

$$\sum_{i=1}^{n} r(l,n)_i U(l,n)_{i,j} = 1. \qquad (6.30)$$

Therefore,

$$(\mathbf{m}(l,n)U(l,n))_j \le f_{l+j}^{1/2}. \qquad (6.31)$$

Using this and (6.16) we get (6.25). $\qquad\qquad\qquad\qquad\qquad\qquad\qquad\qquad\qquad\qquad\square$

We can now give a concrete corollary of Lemma 6.2.

Theorem 6.1 *For any* $\alpha > 0$, *let* $\widetilde{X}_\alpha(l,n) = (\widetilde{X}_{\alpha,l+1},\dots,\widetilde{X}_{\alpha,l+n})$ *be an* α-*permanental random variable determined by the kernel*

$$\{U(l,n)_{j,k} + f_{l+k}\}_{j,k=1}^{n}. \qquad (6.32)$$

Let $\widetilde{Y}_\alpha(l,n) = (\widetilde{Y}_{\alpha,l+1},\dots,\widetilde{Y}_{\alpha,l+n})$ *be an* α-*permanental random variable determined by the symmetric kernel,*

$$\{U(l,n)_{j,k} + a(l,n)_j a(l,n)_k\}_{j,k=1}^{n}, \qquad (6.33)$$

where $a(l,n)_j$, $j = 1,\dots,n$, *is given in (6.25).*
  Suppose that

$$\rho_{l,n} = \sum_{j,k=1}^{n} (U(l,n))_{j,k}^{-1} f_{l+k} \le \delta_l, \quad \text{where} \quad \delta_l = o(l). \qquad (6.34)$$

Then for all functions g *of* $\widetilde{X}_\alpha(l,n)$ *and* $\widetilde{Y}_\alpha(l,n)$, *and sets* $\mathcal{B}$ *in the range of* g, *and all* l *sufficiently large,*

$$P\left(g(\widetilde{Y}_\alpha(l,n)) \in \mathcal{B}\right) - 2\alpha\delta_l \le P\left(g(\widetilde{X}_\alpha(l,n)) \in \mathcal{B}\right) \qquad (6.35)$$

$$\le 2\alpha\delta_l + P\left(g(\widetilde{Y}_\alpha(l,n)) \in \mathcal{B}\right).$$

Proof This follows from Lemma 6.2 and Lemmas 6.3 and 6.4, with $K(l,n+1)$ as defined in (6.8). However we take g in Lemma 6.2 restricted to

$(\widetilde{Y}_{\alpha,1}, \ldots, \widetilde{Y}_{\alpha,n})$ and $(\widetilde{X}_{\alpha,1}, \ldots, \widetilde{X}_{\alpha,n})$. We also use the inequality

$$\left(\frac{1}{1+\rho_{l,n}}\right)^{\alpha} > 1 - 2\alpha\delta_l, \tag{6.36}$$

all l sufficiently large. $\square$

Proof of Theorem 1.2 This is a direct application of Theorem 6.1. We continue with the notation in Theorem 6.1 but initially we restrict ourselves to the cases where $\alpha = k/2$, for integers $k \geq 1$. We use (6.35) with the event

$$\{g(\widetilde{Y}_{k/2}(l,n)) \in \mathcal{B}\} = \left\{ \sup_{1 \leq j \leq n} \frac{\widetilde{Y}_{k/2,l+j}}{\phi_{l+j}} \leq 1 \right\}, \tag{6.37}$$

and similarly for $\{\widetilde{X}_{k/2,l+j}\}_{j=1}^{n}$. We have that for all l sufficiently large and $M > 0$,

$$P\left(\sup_{1 \leq j \leq n} \frac{\widetilde{Y}_{k/2,l+j}}{\phi_{l+j}} \leq M \right) - k\delta_l \leq P\left(\sup_{1 \leq j \leq n} \frac{\widetilde{X}_{k/2,l+j}}{\phi_{l+j}} \leq M \right) \tag{6.38}$$

$$\leq k\delta_l + P\left(\sup_{1 \leq j \leq n} \frac{\widetilde{Y}_{k/2,l+j}}{\phi_{l+j}} \leq M \right).$$

The key point here is that

$$\{\widetilde{Y}_{k/2,l+j}\}_{j=1}^{n} \overset{law}{=} \left\{ \sum_{i=1}^{k} \frac{(\eta_{i,l+j} + a(l,n)_j \xi_i)^2}{2} \right\}_{j=1}^{n}, \tag{6.39}$$

where $\{\eta_{i,l+j} + a(l,n)_j \xi_i\}_{j=1}^{n}$, $i = 1, \ldots, k$, are independent copies of $\{\eta_{l+j} + a(l,n)_j \xi\}_{j=1}^{n}$. This follows from the definition of permanental processes in (1.11).
    We write

$$\sum_{i=1}^{k}(\eta_{i,l+j} + a(l,n)_j \xi_i)^2 = \sum_{i=1}^{k} \eta_{i,l+j}^2 + 2a(l,n)_j \sum_{i=1}^{k} \eta_{i,l+j}\xi_i + a^2(l,n)_j \sum_{i=1}^{k} \xi_i^2$$

$$\leq \sum_{i=1}^{k} \eta_{i,l+j}^2 + 2f_{l+j}^{1/2}\left(\sum_{i=1}^{k} \eta_{i,l+j}^2\right)^{1/2}\left(\sum_{i=1}^{k} \xi_i^2\right)^{1/2} + f_{l+j}\sum_{i=1}^{k} \xi_i^2, \tag{6.40}$$

by (6.25). Therefore,

$$\sup_{1 \le j \le n} \frac{\widetilde{Y}_{k/2,l+j}}{2\phi_{l+j}} \tag{6.41}$$

$$\le \sup_{1 \le j \le n} \frac{\sum_{i=1}^{k} \eta_{i,l+j}^2}{2\phi_{l+j}} + 2\epsilon_l^{1/2} \left(\rho_k \sup_{1 \le j \le n} \frac{\sum_{i=1}^{k} \eta_{i,l+j}^2}{2\phi_{l+j}} \right)^{1/2} + \epsilon_l \chi_k,$$

$$:= \sup_{1 \le j \le n} \frac{\sum_{i=1}^{k} \eta_{i,l+j}^2}{2\phi_{l+j}} + A_{l,n} + B_l,$$

where $\epsilon_l = \sup_{j \ge 1}(f_{l+j}/\phi_{l+j})$, which by (1.18), goes to zero as $l \to \infty$, and $\chi_k = \sum_{i=1}^{k} \xi_i^2$.

Consider the first inequality in (6.38) and take the limit as $n \to \infty$. For all $\epsilon > 0$ we have,

$$P\left(\sup_{1 \le j \le \infty} \frac{\widetilde{X}_{k/2,l+j}}{\phi_{l+j}} \le 1 + \epsilon \right) \tag{6.42}$$

$$\ge P\left(\sup_{1 \le j \le \infty} \frac{\sum_{i=1}^{k} \eta_{i,l+j}^2}{2\phi_{l+j}} \le 1 + \epsilon - A_{l,\infty} - B_l \right) - k\delta_l.$$

Similarly, it follows from the second inequality in (6.38) and the analogue of (6.41) for the lower bound, that for all $\epsilon > 0$ we have,

$$P\left(\sup_{1 \le j \le \infty} \frac{\widetilde{X}_{k/2,l+j}}{\phi_{l+j}} \le 1 - \epsilon \right) \tag{6.43}$$

$$\le P\left(\sup_{1 \le j \le \infty} \frac{\sum_{i=1}^{k} \eta_{i,l+j}^2}{2\phi_{l+j}} \le 1 - \epsilon + A_{l,\infty} + B_l \right) + k\delta_l.$$

It follows from (1.17) and Lemma 6.5 below that,

$$\lim_{j \to \infty} \frac{\sum_{i=1}^{k} \eta_{i,j}^2}{2\phi_j} = 1, \qquad a.s. \tag{6.44}$$

Therefore, if we take the limits in (6.42) and (6.43) as $l \to \infty$ we get that for all $\epsilon' > 0$.

$$1 - \epsilon' \le \varlimsup_{j \to \infty} \frac{\widetilde{X}_{k/2,j}}{\phi_j} \le 1 + \epsilon', \qquad a.s. \tag{6.45}$$

and since this holds for all $\epsilon > 0$ we get,

$$\varlimsup_{j \to \infty} \frac{\widetilde{X}_{k/2,j}}{\phi_j} = 1, \qquad a.s. \tag{6.46}$$

Now, suppose that $1/2 \leq \alpha \leq k'$ for some integer k'. Since (6.46) holds for $k = 1/2$ and $k = k'$ we can use the property that α-permanental processes are infinitely divisible and positive to see that (1.19) holds.   □

Lemma 6.5 *Let $\{\eta_j; j \in \overline{\mathbb{N}}\}$ be a Gaussian sequence and for each $i \in \overline{\mathbb{N}}$, let $\{\eta_{i,j}, j \in \overline{\mathbb{N}}\}$ be an independent copy of $\{\eta_j; j \in \overline{\mathbb{N}}\}$. Let $\{\phi_j\}$ be a sequence such that,*

$$\overline{\lim_{j \to \infty}} \frac{|\eta_j|}{(2\phi_j)^{1/2}} = 1 \quad a.s., \tag{6.47}$$

then for any integer $k \geq 1$,

$$\overline{\lim_{j \to \infty}} \frac{\sum_{i=1}^{k} \eta_{i,j}^2}{2\phi_j} = 1, \quad a.s. \tag{6.48}$$

This also holds for a Gaussian process $\{\eta_t; t \in R^+\}$.

Proof We follow the proof of the law of the iterated logarithm for Brownian motion in [19, Theorem 18.1]. Clearly, we only need to prove the upper bound.

Fix k and $\epsilon > 0$ and $X_j = (\eta_{1,j}, \ldots, \eta_{k,j})$ and u be a unit vector in R^k. By checking their covariances we see that

$$\{(u \cdot X_j), \ j = 1, \ldots\} \stackrel{law}{=} \{\eta_j, \ j = 1, \ldots\}. \tag{6.49}$$

Therefore, by (6.47),

$$\overline{\lim_{j \to \infty}} \frac{|(u \cdot X_j)|}{(2\phi_j)^{1/2}} = 1 \quad a.s. \tag{6.50}$$

Note that $\|X_j\|_2 = (\sum_{i=1}^{k} \eta_{i,j}^2)^{1/2}$ and,

$$\overline{\lim_{j \to \infty}} \frac{\|X_j\|_2}{(2\phi_j)^{1/2}} = \overline{\lim_{j \to \infty}} \sup_{\|u\|_2=1} \frac{|(u \cdot X_j)|}{(2\phi_j)^{1/2}}. \tag{6.51}$$

For any $\epsilon > 0$ we can find a finite set of unit vectors $\mathcal{U}(m) = (u_1, \ldots, u_m\}$ in R^k with the property that for any unit vector u in R^k, $\inf_{1 \leq l \leq m} \|u - u_l\|_1 \leq \epsilon/k$.

Let u be a unit vector in R^k. For all $u_l \in \mathcal{U}(m)$,

$$|(u \cdot X_j)| \leq |((u - u_l) \cdot X_j)| + |(u_l \cdot X_j)| \tag{6.52}$$

$$\leq \|u - u_l\|_1 \left| \frac{(u - u_l)}{\|u - u_l\|_1} \cdot X_j \right| + \sup_{l=1,\ldots,M} |(u_l \cdot X_j)|$$

$$\leq \|u - u_l\|_1 \left| \frac{(u - u_l)}{\|u - u_l\|_1} \cdot X_j \right| + \sup_{l=1,\ldots,M} |(u_l \cdot X_j)|$$

$$\leq \|u - u_l\|_1 \sum_{i=1}^{k} |\eta_{i,j}| + \sup_{l=1,\ldots,M} |(u_l \cdot X_j)|.$$

Since this holds for all $u_l \in \mathcal{U}(m)$, we see that,

$$|(u \cdot X_j)| \leq \min_{1 \leq l \leq m} \|u - u_l\|_1 \sum_{i=1}^{k} |\eta_{i,j}| + \sup_{l=1,\ldots,M} |(u_l \cdot X_j)| \qquad (6.53)$$

$$\leq \frac{\epsilon}{k} \sum_{i=1}^{k} |\eta_{i,j}| + \sup_{l=1,\ldots,M} |(u_l \cdot X_j)|.$$

Consequently,

$$\varlimsup_{j \to \infty} \sup_{\|u\|_2=1} \frac{|(u \cdot X_j)|}{(2\phi_j)^{1/2}} \leq \varlimsup_{j \to \infty} \sup_{l=1,\ldots,M} \frac{|u_l \cdot X_j|}{(2\phi_j)^{1/2}} + \frac{\epsilon}{k} \sum_{i=1}^{k} \varlimsup_{j \to \infty} \frac{|\eta_{i,j}|}{(2\phi_j)^{1/2}}. \qquad (6.54)$$

It follows from (6.47) that the last term is bounded by ϵ.

Let Ω' be the event that equality holds in (6.50) with $u = u_l$ for all $l = 1, \ldots, m$. It follows that for any ϵ and any $\omega \in \Omega'$ we can find $j_0(\omega)$ such that

$$\frac{|u_l \cdot X_j(\omega)|}{(2\phi_j)^{1/2}} \leq 1 + \epsilon, \qquad \forall j \geq j_0(\omega) \text{ and all } 1 \leq l \leq m. \qquad (6.55)$$

Since $P(\Omega') = 1$, it now follows from (6.54) that

$$\varlimsup_{j \to \infty} \sup_{\|u\|_2=1} \frac{|(u \cdot X_j)|}{(2\phi_j)^{1/2}} \leq 1 + 2\epsilon, \qquad a.s. \qquad (6.56)$$

Since this holds for all $\epsilon > 0$, the upper bound for (6.48) follows from (6.51). □

Chapter 7
Permanental Sequences with Kernels That Have Uniformly Bounded Row Sums

When the kernel of a permanental sequence has uniformly bounded row sums we can obtain its rate of growth at infinity without using the difficult machinery of Chap. 6. Let M be an $\overline{\mathbb{N}} \times \overline{\mathbb{N}}$ matrix and consider the operator norm on $\ell_\infty \to \ell_\infty$,

$$\|M\| = \sup_{\|x\|_\infty \le 1} \|Mx\|_\infty = \sup_j \sum_k |M_{j,k}|. \tag{7.1}$$

Lemma 7.1 *Let* $M = \{M_{j,k}, \; j, k \in \overline{\mathbb{N}}\}$ *be a positive matrix and assume that both* $\|M\|$ *and* $\|M^T\| < \infty$. *Then for all* $\epsilon > 0$, *there exists a sequence* $\{i_n, n \in \overline{\mathbb{N}}\}$ *such that* $i_n \le n(\|M\| + \|M^T\|)/\epsilon$, *for all* $n \in \overline{\mathbb{N}}$, *and*

$$M_{i_j, i_k} \le \epsilon, \qquad \forall \, j, k \in \overline{\mathbb{N}}, \quad j \ne k. \tag{7.2}$$

Proof Assume to begin that M is symmetric. Fix $\epsilon > 0$, and consider $\{M_{1,k}\}_{k=1}^\infty$. Not more than $\|M\|/\epsilon$ of these terms can be greater than ϵ. Let $\{M_{1,k_1(p_i)}, i = 1, \ldots C_1\}$ denote the terms in $\{M_{1,k}\}_{k=1}^\infty$ which are greater than ϵ and set $R_1 = \{k_1(p_i), i = 1, \ldots C_1\}$. As we just pointed out $|R_1| \le \|M\|/\epsilon$.

Note that

$$M_{1,k} \le \epsilon \qquad \forall \, k \in R_1^c. \tag{7.3}$$

Set $i_1 = 1$ and set i_2 equal to the smallest index in R_1^c that is greater than i_1.

We repeat this procedure starting with $M_{i_2,k}$ with $k \in R_1^c$ to get R_2 where $|R_2| \le \|M\|/\epsilon$ and,

$$M_{i_2,k} \le \epsilon, \qquad \forall \, k \in R_2^c. \tag{7.4}$$

© The Author(s), under exclusive license to Springer Nature Switzerland AG 2021
M. B. Marcus and J. Rosen, *Asymptotic Properties of Permanental Sequences*,
SpringerBriefs in Probability and Mathematical Statistics,
https://doi.org/10.1007/978-3-030-69485-2_7

Therefore, for $j = 1, 2$,

$$M_{i_j,k} \leq \epsilon \qquad \forall k \in (R_1 \cup R_2)^c = R_1^c \cap R_2^c. \tag{7.5}$$

We continue this procedure setting i_3 equal to the smallest integer in $(R_1 \cup R_2)^c$ that is greater than i_2, and so on, to get $\{i_n, n \in \overline{\mathbb{N}}\}$. This completes the proof when M is symmetric.

More generally, assume only that both $\|M\|$ and $\|M^T\| < \infty$. We use a construction similar to the one above but we work alternately with both M and M^T. Therefore, we can obtain $\{i_l, l = 1, \ldots, n\}$ and a set $S_n \subset \overline{\mathbb{N}}$ such that $|S_n| \leq n(\|M\| + \|M^T\|)/\epsilon$ and for $l = 1, \ldots, n$,

$$M_{i_l,k} \leq \epsilon, \quad \text{and} \quad M_{k,i_l} \leq \epsilon, \qquad \forall k \in (S_n)^c. \tag{7.6}$$

Choose i_{n+1} equal to the smallest integer in $(S_n)^c$ that is greater than i_n.

We continue the above procedure starting with $M_{i_{n+1},k}$ and $M_{k,i_{n+1}}$, with $k \in S_n^c$, to get R_{n+1} where $|R_{n+1}| \leq (\|M\| + \|M^T\|)/\epsilon$ and,

$$M_{i_{n+1},k} \leq \epsilon, \quad \text{and} \quad M_{k,i_{n+1}} \leq \epsilon, \qquad \forall k \in R_{n+1}^c. \tag{7.7}$$

Therefore, for $j = 1, \ldots, n+1$,

$$M_{i_j,k} \leq \epsilon, \quad M_{k,i_j} \leq \epsilon \qquad \forall k \in (S_n \cup R_{n+1})^c = S_n^c \cap R_{n+1}^c. \tag{7.8}$$

This shows that for all $\epsilon > 0$, there exists a sequence $\{i_n, n \in \overline{\mathbb{N}}\}$ such that $i_n \leq n(\|M\| + \|M^T\|)/\epsilon$, for all $n \in \overline{\mathbb{N}}$, and in particular that (7.2) holds. $\qquad\square$

Proof of Theorem 1.3 It follows from (1.11) that for all $j \in \overline{\mathbb{N}}$,

$$\frac{\widetilde{X}_{\alpha,j}}{\widetilde{U}_{j,j}} \overset{law}{=} \xi_\alpha, \tag{7.9}$$

where ξ_α has probability density function $x^{\alpha-1}e^{-x}/|\Gamma(\alpha)$. Using the Borel-Cantelli Lemma, we get,

$$\limsup_{n \to \infty} \frac{\widetilde{X}_{\alpha,n}}{\widetilde{U}_{n,n} \log n} \leq 1 \qquad a.s. \tag{7.10}$$

This gives the upper bound in (1.21) because since $f \in c_0^+$ and $\inf U_{n,n} > 0$, we have

$$\lim_{n \to \infty} \frac{U_{n,n}}{\widetilde{U}_{n,n}} = 1. \tag{7.11}$$

To get the lower bound in (1.21) consider,

$$\widehat{U}_{j,k} = \frac{\widetilde{U}_{j,k}}{(\widetilde{U}_{j,j}\widetilde{U}_{k,k})^{1/2}}, \qquad j,k \in \overline{\mathbb{N}}. \tag{7.12}$$

It follows from Lemma 7.1 that for all $\epsilon > 0$ there exists a sequence $\{i_n, n \in \overline{\mathbb{N}}\}$ with

$$i_n \leq 2n\|U\|/(\epsilon\delta), \qquad \forall n \in \overline{\mathbb{N}}, \tag{7.13}$$

such that,

$$U_{i_j,i_k} \leq \frac{\epsilon\delta}{2}, \qquad \forall j,k \in \overline{\mathbb{N}}, \quad j \neq k, \tag{7.14}$$

Therefore,

$$\widehat{U}_{i_j,i_k} + \widehat{U}_{i_k,i_j} \leq \frac{2U_{i_j,i_k} + f(i_k) + f(i_j)}{(U_{i_j,i_j}U_{i_k,i_k})^{1/2}} \leq \epsilon + \frac{f(i_k) + f(i_j)}{\delta}. \tag{7.15}$$

Using the fact that $f \in c_0^+$ we see that we can find an n_0 such that

$$\widehat{U}_{i_j,i_k} + \widehat{U}_{i_k,i_j} \leq 2\epsilon, \qquad \forall j,k \geq n_0. \tag{7.16}$$

Therefore, by [15, Lemma 7.1],

$$\limsup_{n\to\infty} \frac{\widetilde{X}_{\alpha,i_n}}{\widetilde{U}_{i_n,i_n}\log(n-n_0)} \geq 1 - 6\epsilon \qquad a.s., \tag{7.17}$$

or, equivalently,

$$\limsup_{n\to\infty} \frac{\widetilde{X}_{\alpha,i_n}}{\widetilde{U}_{i_n,i_n}\log n} \geq 1 - 6\epsilon \qquad a.s. \tag{7.18}$$

Using (7.11) and (7.13), we get,

$$\limsup_{n\to\infty} \frac{\widetilde{X}_{\alpha,i_n}}{U_{i_n,i_n}\log i_n} \geq 1 - 6\epsilon \qquad a.s. \tag{7.19}$$

which gives (1.21). □

If $f = Uh$, where $\|U\| < \infty$, and $h \in \ell_1$ then since $\sum_i f_i = \sum_i \sum_j U_{i,j}h_j$, it follows using the symmetry of U that $f \in \ell_1^+$ and consequently in c_0^+. However, when U has some regularity, $f \in c_0^+$ if and only if $h \in c_0^+$.

Lemma 7.2 *Let $f = Uh$, where $\|U\| < \infty$, $h \in c_0^+$ and there exists a sequence $\{k_j\}$, $\lim_{j\to\infty} k_j = \infty$, such that*

$$\lim_{j\to\infty} \sum_{k=1}^{k_j} U_{j,k} = 0. \tag{7.20}$$

Then $f \in c_0^+$.

If $\inf_j U_{j,j} > 0$ then $f \in c_0^+$ implies that $h \in c_0^+$.

Proof The first statement follows from the inequality,

$$f_j = \sum_{k=1}^{\infty} U_{j,k} h_k \leq \|h\|_{\infty} \sum_{k=1}^{k_j} U_{j,k} + \|U\| \sup_{k \geq k_j} |h_k|. \tag{7.21}$$

The second statement is obvious, since,

$$f_j = \sum_{k=1}^{\infty} U_{j,k} h_k \geq U_{j,j} h_j. \tag{7.22}$$

$\square$

Proof of Theorem 1.4 We show in [15, Theorem 6.1] that $\widetilde{U}$ is the kernel of an α-permanental sequence. Also, it follows from (1.23) that $f \in \ell_1^+ \subset c_0^+$ and that the hypotheses of Lemma 7.1 are satisfied. Consequently, the proof follows as in the proof of Theorem 1.3. $\square$

Proof of Theorem 1.11 The Lévy process X is obtained by killing a Lévy process say $\widehat{X}$ on $\mathbb{Z}$ at the end of an independent exponential time with mean $1/\beta$. Let $\{\widehat{p}_t(i, j); j, k \in \mathbb{Z}\}$ denote the transition densities for $\widehat{X}$ and $\{p_t(i, j); j, k \in \mathbb{Z}\}$ the transition densities for X. We have

$$p_t(i, j) = e^{-\beta t} \widehat{p}_t(i, j). \tag{7.23}$$

Consequently,

$$U_{j,k} = \int_0^{\infty} e^{-\beta t} \widehat{p}_t(j, k) \, dt, \qquad \forall j, k \in \mathbb{Z}. \tag{7.24}$$

Since $\widehat{X}$ is a Levy process we have

$$\widehat{p}_t(i, j) = \widehat{p}_t(0, j - i) := \widehat{p}_t(j - i). \tag{7.25}$$

Therefore, for all $j \in \mathbb{Z}$,

$$U_{j,j} = U_{0,0} = \int_0^{\infty} e^{-\beta t} \widehat{p}_t(0) \, dt = \int_0^{\infty} p_t(0) \, dt. \tag{7.26}$$

To see that (1.66) is the kernel of an α-permanental sequence we first note that since X is an exponentially killed Lévy process on $\mathbb{Z}$ with potential U, then $\overline{X} = -X$ is also a Lévy process on $\mathbb{Z}$, the dual of X, with transition densities

$$\overline{p}_t(i, j) = p_t(-i, -j) := p_t(i - j) = p_t(j, i), \tag{7.27}$$

and consequently, potential

$$\overline{U}_{i,j} = U_{j,i}, \quad i, j \in \mathbb{Z}. \tag{7.28}$$

The proof that (1.66) is the kernel of an α-permanental process for all functions f that are finite excessive functions for X proceeds in three steps.

We first show that for any $g = \{g_k\}$ where $g_k = \sum_{j=-\infty}^{\infty} \overline{U}_{k,j} h_j$, and $h \in \ell_1^+(\mathbb{Z})$,

$$U_{j,k} + g_k, \quad j, k \in \mathbb{Z}, \tag{7.29}$$

is the kernel of an α-permanental process. To see this note that by (7.28), $g_k = \sum_{j=-\infty}^{\infty} \overline{U}_{k,j} h_j = \sum_{j=-\infty}^{\infty} h_j U_{j,k}$. Therefore it follows from [15, Theorem 6.1] that (7.29) is the restriction to $\mathbb{Z} \times \mathbb{Z}$ of the potential of a transient Borel right process $\widetilde{X} = (\Omega, \mathcal{F}_t, \widetilde{X}_t, \theta_t, \widetilde{P}^x)$ with state space $\mathbb{Z} \cup \{*\}$, where $*$ is an isolated point. Consequently, (7.29) is the kernel of an α-permanental sequence.

We show next that (7.29) is the kernel of an α-permanental process for any g that is a finite excessive function for $\overline{X}$. We use the following lemma which is Lemma 6.2 in [15].

Lemma 7.3 *Assume that for each $n \in \mathbb{N}$, $u^{(n)}(s, t)$, $s, t \in S$, is the kernel of an α-permanental process. If $u^{(n)}(s, t) \to u(s, t)$ for all $s, t \in S$, then $u(s, t)$ is the kernel of an α-permanental process.*

We now use arguments from the proof of [15, Theorem 1.11]. Consider a general function $g = \{g_k\}$ that is a finite excessive function for $\overline{X}$. It follows from [2, II, (2.19)] that there exists a sequence of functions $h^{(n)} = \{h_k^{(n)}\} \in \ell_\infty^+(\mathbb{Z})$ such that $g^{(n)}$ defined by,

$$g_k^{(n)} = \sum_{j=-\infty}^{\infty} \overline{U}_{k,j} h_j^{(n)}, \tag{7.30}$$

is also in $\ell_\infty^+(\mathbb{Z})$ and is such that for each $k \in \overline{\mathbb{N}}$, $g_k^{(n)} \uparrow g_k$.

If $h^{(n)} \in \ell_1^+$ then by the first step in this proof we have that $\{U_{j,k} + g_k^{(n)}\}$, $j, k \in \mathbb{Z}\}$ are kernels of α-permanental processes. Consequently, by Lemma 7.3, (7.29) is the kernel of α-permanental process.

If $h^{(n)} \notin \ell_1^+$ we first consider $h^{(n)} 1_{[-m,m]}$ which clearly is in ℓ_1^+ for each $m < \infty$. We then set

$$g_k^{(n,m)} = \sum_{j=-\infty}^{\infty} \overline{U}_{k,j} h_j^{(n)} 1_{\{-m \le j \le m\}}. \tag{7.31}$$

Therefore, as in the previous paragraph, we have that $\{U_{j,k} + g_k^{(n,m)}, j, k \in \mathbb{Z}\}$ is the kernel of an α-permanental process. Taking the limit as $m \to \infty$, it follows from Lemma 7.3 that $\{U_{j,k} + g_k^{(n)}, j, k \in \mathbb{Z}\}$ is the kernel of an α-permanental process. Since $g_k^{(n)} \to g_k$ we use Lemma 7.3 again to see that (7.29) is the kernel of an α-permanental process for all finite excessive functions g for $\overline{X}$.

The last step in the proof that (1.66) is the kernel of an α-permanental process is to show that when f_k is a finite excessive function for X, then f_{-k} is a finite excessive function for $\overline{X}$. To see this, note that if f_k is a finite excessive function for X, then, by definition,

$$\sum_{k=-\infty}^{\infty} p_t(k-i) f_k = \sum_{k=-\infty}^{\infty} p_t(i,k) f_k \uparrow f_i, \qquad \text{as } t \downarrow 0. \tag{7.32}$$

It follows from this that as $t \downarrow 0$,

$$\sum_{k=-\infty}^{\infty} p_t(k,i) f_{-k} = \sum_{k=-\infty}^{\infty} p_t(-k-(-i)) f_{-k} \uparrow f_{-i}. \tag{7.33}$$

Consequently f_{-k} is a finite excessive function for $\overline{X}$.

This completes the proof that (1.66) is the kernel of an α-permanental process. Using this and the fact that $\lim_{k\to\infty} f_{-k} = 0$, proceeding exactly as in the proof of Theorem 1.3, we get the upper bound in (1.67).

To obtain the lower bound in (1.67) we note that by (7.24) (7.25) and Fubini's Theorem,

$$\sum_{i=-\infty}^{\infty} U_{i,j} = \sum_{j=-\infty}^{\infty} U_{i,j} = \int_0^\infty e^{-\beta t}\, dt = \frac{1}{\beta}. \tag{7.34}$$

Using this and (7.26) we see that the conditions in (1.23) and Lemma 7.1 are all satisfied for $\{U_{j,k}; j, k \in \overline{\mathbb{N}}\}$. Therefore, as in the proofs of Theorems 1.3 and 1.4, the lower bound in (1.67) follows from [15, Lemma 7.1].

To verify the last statement in this theorem we see from the proof of Lemma 7.2, that we need only show that there exists a sequence $\{k_j\}$, $\lim_{j\to\infty} k_j = \infty$, such that

$$\lim_{|j|\to\infty} \sum_{k=-k_{|j|}}^{k_{|j|}} U_{j,k} = 0. \tag{7.35}$$

We have,

$$\lim_{|j|\to\infty} \sum_{k=-k_{|j|}}^{k_{|j|}} U_{j,k} = \lim_{|j|\to\infty} \sum_{k=-k_{|j|}}^{k_{|j|}} U_{0,-j+k} = \lim_{|j|\to\infty} \sum_{l=-j-k_{|j|}}^{-j+k_{|j|}} U_{0,l}. \tag{7.36}$$

It follows from (7.34) that this last term goes to zero when $k_j = j/2$. $\qquad \square$

Chapter 8
Uniform Markov Chains

Lemma 8.1 *Let $X = (\Omega, \mathcal{F}_t, X_t, \theta_t, P^x)$ be a transient Borel right process with state space $\overline{\mathbb{N}}$, finite Q-matrix Q, and strictly positive potential $U = \{U_{j,k}, j, k \in \overline{\mathbb{N}}\}$. Then,*

$$- \delta_{i,l} = \sum_{j=1}^{\infty} Q_{i,j} U_{j,l}, \quad \text{for all } i, l \in \overline{\mathbb{N}}. \tag{8.1}$$

Proof Set $q(i) = -Q_{i,i}$. Without loss of generality we can take $q(i) > 0$. For any function h we have,

$$U h(i) = \frac{h(i)}{q(i)} + \sum_{j=1, j\neq i}^{\infty} \frac{Q_{i,j}}{q(i)} U h(j). \tag{8.2}$$

To see this, let τ_i be the time of the first exit from state i and note that,

$$U h(i) = E^i \left(\int_0^{\tau_i} h(X_t) \, dt \right) + E^{X_{\tau_i}} \left(\int_0^{\infty} h(X_t) \, dt \right). \tag{8.3}$$

Using the facts that the exit time is an exponential random variable with expectation $1/q(i)$, and the probability that upon exit the process jumps from i to j is $Q_{i,j}/q(i)$, we get the two terms in (8.2).

It follows from (8.2) that

$$- h(i) = -q(i) U h(i) + \sum_{j=1, j\neq i}^{\infty} Q_{i,j} U h(j) = \sum_{j=1}^{\infty} Q_{i,j} U h(j). \tag{8.4}$$

Take $h(k) = \delta_{l,k}$ and note that,

© The Author(s), under exclusive license to Springer Nature Switzerland AG 2021
M. B. Marcus and J. Rosen, *Asymptotic Properties of Permanental Sequences*,
SpringerBriefs in Probability and Mathematical Statistics,
https://doi.org/10.1007/978-3-030-69485-2_8

$$\sum_{j=1}^{\infty} Q_{i,j} U h(j) = \sum_{j=1}^{\infty} Q_{i,j} \sum_{k=1}^{\infty} U_{j,k} \delta_{l,k} = \sum_{j=1}^{\infty} Q_{i,j} U_{j,l}. \qquad (8.5)$$

Therefore, by (8.4),

$$- \delta_{l,i} = \sum_{j=1}^{\infty} Q_{i,j} U_{j,l}, \qquad (8.6)$$

which is (8.1).   □

Lemma 8.1 gives the following useful inequality:

Lemma 8.2 *Let X, Q and U be as defined in Lemma 8.1. Then,*

$$U_{i,i} \geq \frac{1}{|Q_{i,i}|}, \qquad \forall i \in \overline{\mathbb{N}}. \qquad (8.7)$$

Proof Since $Q(i,i) < 0$ it follows from (8.1) that

$$1 = |Q_{i,i}| U_{i,i} - \sum_{j \neq i} Q_{i,j} U_{j,i}, \qquad (8.8)$$

and since $Q(i, j) \geq 0$ for $i \neq j$ we get (8.7).   □

The inequality in (8.7) can also be obtained from the facts that $1/|Q_{i,i}| = 1/q(i)$ is the expected amount of time the process spends at i during each visit to i, whereas $U_{i,i}$ is the total expected amount of time spent at i when the process starts at i.

We say that a Markov chain X is uniform when it's Q matrix has the property that $\|Q\| < \infty$. When a Markov chain is uniform we can give additional relationships between it's Q matrix and its potential. Since all the row sums of Q are negative,

$$\sup_{j} |Q_{j,j}| \leq \|Q\| \leq 2 \sup_{j} |Q_{j,j}|. \qquad (8.9)$$

Lemma 8.3 *Let X, Q and U be as defined in Lemma 8.1 and assume that X is a uniform Markov chain.*

(i) If the row sums of Q are bounded away from 0 then $\|U\| < \infty$.
(ii) If in addition if Q is a $(2m + 1)$−diagonal matrix for some $m \geq 1$,

$$U_{i,k} \leq C e^{-\lambda |i-k|}, \qquad \forall i, k \in \overline{\mathbb{N}}, \qquad (8.10)$$

for some constants $C, \lambda > 0$.

Proof (i) If $\|Q\| < \infty$ and the row sums of Q are bounded away from 0, then there exists $\beta > 0$ such that ,

$$\delta = \|I + \beta Q\| < 1. \qquad (8.11)$$

It then follows from [7, Sect. 5.3] that $\|e^{t(I+\beta Q)}\| \leq e^{\delta t}$, or equivalently, $\|e^{tQ}\| \leq e^{-(1-\delta)t/\beta}$. Using [7, Sect. 5.3] again, and the fact that the transition semi-group, $P_t = e^{tQ}$, we have

$$\left\| \int_0^\infty P_t \, dt \right\| < \infty. \tag{8.12}$$

Since $U = \int_0^\infty P_t \, dt$, we have $\|U\| < \infty$.

(ii) Let $\sigma = \inf\{t \mid X_t \neq X_0\}$, the time of the first jump of X. Then for all $n \in \overline{\mathbb{N}}$,

$$P^n\left(X_\sigma \in \overline{\mathbb{N}}\right) = \sum_{i \neq n} P^n\left(X_\sigma = i\right) = \sum_{i \neq n} \frac{Q_{n,i}}{|Q_{n,n}|}. \tag{8.13}$$

Note that since the row sums of Q are bounded away from 0 there exists a $\delta > 0$ such that,

$$|Q_{n,n}| - \sum_{i \neq n} Q_{n,i} \geq \delta, \tag{8.14}$$

uniformly in n. Furthermore, since $\sup_n |Q_{n,n}| \leq \|Q\|$, we have,

$$\sum_{i \neq n} \frac{Q_{n,i}}{|Q_{n,n}|} \leq 1 - \frac{\delta}{\|Q\|}. \tag{8.15}$$

Therefore, by (8.13),

$$P^n\left(X_\sigma \in \overline{\mathbb{N}}\right) \leq 1 - \frac{\delta}{\|Q\|}, \qquad \forall n \in \overline{\mathbb{N}}. \tag{8.16}$$

We show immediately below that for all $i < k, i, k \in \overline{\mathbb{N}}$,

$$P^i\left(T_k < \infty\right) \leq e\left(1 - \frac{\delta}{\|Q\|}\right)^{(k-i)/m}. \tag{8.17}$$

Since,

$$U_{i,k} = P^i\left(T_k < \infty\right) U_{k,k} \leq P^i\left(T_k < \infty\right) \|U\|, \tag{8.18}$$

it follows that

$$U_{i,k} \leq e\|U\| \left(1 - \frac{\delta}{\|Q\|}\right)^{(k-i)/m}, \qquad i < k. \tag{8.19}$$

This gives (8.10) with $C = e\|U\|$ and $\lambda = \delta/\|Q\|$.

We now obtain (8.17). Let $[(k-i)/m] = l$ and

$$L_j = \{j, j+1, \ldots, j+m-1\}. \tag{8.20}$$

Since the Markov chain X can move at most m units at each jump,

$$\{X_0 = i, T_k < \infty\} = \{X_0 = i\} \cap_{j=1}^{l-1} \{S_j < \infty\} \cap \{T_k \circ S_{l-1} < \infty\}, \qquad (8.21)$$

where $S_1 = T_{L_{i+1}}$ and $S_j = T_{L_{i+(j-1)m+1}} \circ S_{j-1}, j = 2, \dots, l-1$. Then by the Markov property and (8.16)

$$P^i (T_k < \infty) = E^i \left(\cap_{j=1}^{l-1} \{S_j < \infty\} E^{X_{S_{l-1}}} (T_k < \infty) \right) \qquad (8.22)$$

$$\leq \left(1 - \frac{\delta}{\|Q\|} \right) E^i \left(\cap_{j=1}^{l-1} \{S_j < \infty\} \right)$$

$$\leq \left(1 - \frac{\delta}{\|Q\|} \right) E^i \left(\cap_{j=1}^{l-2} \{S_j < \infty\} E^{X_{S_{l-2}}} \left(T_{L_{i+(l-2)m+1}} < \infty \right) \right)$$

$$\leq \left(1 - \frac{\delta}{\|Q\|} \right)^2 E^i \left(\cap_{j=1}^{l-2} \{S_j < \infty\} \right).$$

Continuing this procedure we get

$$P^i (T_k < \infty) \leq \left(1 - \frac{\delta}{\|Q\|} \right)^l, \qquad (8.23)$$

which gives (8.17).   □

Proof of Theorem 1.5 To show that the first condition in (1.20) holds we use Lemma 8.2 and (8.9) to see that,

$$U_{i,i} \geq \frac{1}{|Q_{i,i}|} \geq \frac{1}{\sup_j |Q_{j,j}|} \geq \frac{1}{\|Q\|}. \qquad (8.24)$$

The second condition in (1.20) is given in Lemma 8.3.

Now suppose that Q is a $(2m + 1)$−diagonal matrix for some $m \geq 1$. It follows from (8.10) and Lemma 7.2 that $f \in c_0^+$ if and only if $f = Uh$ for some $h \in c_0^+$. □

Remark 8.1 When X in Theorem 1.1 is a uniform Markov chain with Q-matrix Q and $f = Uh$ with $h \in \ell_1^+$, then it follows from the proof of the theorem that $\widetilde{U}$ is the restriction to $\overline{\mathbb{N}}$ of the potential of a uniform Markov chain on $\{0\} \cup \overline{\mathbb{N}}$ with Q-matrix

$$Q_{j,k}, \qquad j, k \in \overline{\mathbb{N}}, \qquad (8.25)$$

$$Q_{0,0} = 1 + \|h\|_1, \quad Q_{j,0} = - \sum_{k=1}^{\infty} Q_{j,k}, \, j \in \overline{\mathbb{N}}, \quad \text{and} \quad Q_{0,k} = -h_k, k \in \overline{\mathbb{N}}.$$

It is clear that all the row sums of this Q-matrix are equal to 0, except for the first row sum which is equal to 1.

At the ends of Chaps. 3, 4 and 5 we examine the effects on the covariances of certain Gaussian sequences that are also potentials of Markpov chains when we shift a parameter **s** by $\Delta > s_1$. We show that when the 'shifted' covariance is itself a potential, all the elements of the Q matrix of this new potential is are equal to the elements Q matrix of the original potential, except for $(1, 1)$ coordinate with is a function of Δ. (See page 43).) The next lemma reverses and generalizes this proceedure. It examines the effects on the potentials of Markov chains when we change any one term of their Q-matrices. We consider the matrix $E(k, l) = \{E(k, l)_{i,j}; i, j \in \overline{\mathbb{N}}\}$, with one non-zero element, where

$$E(k, l)_{i,j} = \delta_{(k,l)}(i, j). \tag{8.26}$$

Lemma 8.4 *Let Q be the Q-matrix of a symmetric transient uniform Markov chain on $\overline{\mathbb{N}}$ with potential U satisfying,*

$$0 < U_{j,k}Q_{j,k} < U_{k,k}Q_{j,k}, \quad \text{for some } j \neq k \in \overline{\mathbb{N}}, \tag{8.27}$$

and assume that for some real number b the matrix,

$$Q + bE(k, l), \tag{8.28}$$

is the Q-matrix of a transient Markov chain X on $\overline{\mathbb{N}}$.
 Then if either,

(i) Q is a $(2n + 1)-$diagonal matrix for some $n \geq 1$,
 or
(ii) $\sum_{j=1}^{\infty} U_{i,j} < \infty$ for each $i \geq 1$,

we have, $b < 1/U_{l,k}$ and the potential of X is given by $W = \{W_{i,j}; i, j \in \overline{\mathbb{N}}\}$ where,

$$W_{i,j} = U_{i,j} + \frac{bU_{i,k}U_{l,j}}{1 - bU_{k,l}}. \tag{8.29}$$

Proof In order for (8.28) to be the Q-matrix of a transient Markov chain X on $\overline{\mathbb{N}}$, we must have,

$$Q_{k,k} + b + \sum_{j=1, j\neq k}^{\infty} Q_{k,j} \leq 0. \tag{8.30}$$

Therefore,

$$Q_{k,k}U_{k,k} + bU_{k,k} + U_{k,k}\sum_{j=1, j\neq k}^{\infty} Q_{k,j} \leq 0. \tag{8.31}$$

It follows from (8.27) that

$$\sum_{j=1,j\neq k}^{\infty} Q_{k,j} U_{j,k} < U_{k,k} \sum_{j=1,j\neq k}^{\infty} Q_{k,j}. \tag{8.32}$$

Consequently,

$$Q_{k,k} U_{k,k} + b U_{k,k} + \sum_{j=1,j\neq k}^{\infty} Q_{k,j} U_{j,k} < 0. \tag{8.33}$$

Therefore, by Lemma 8.1,

$$-1 + b U_{k,k} < 0, \tag{8.34}$$

or, equivalently,

$$b < \frac{1}{U_{k,k}}. \tag{8.35}$$

Since $U_{l,k} \leq U_{k,k}$, this implies that $b < 1/U_{l,k}$ for all $l \in \overline{\mathbb{N}}$. We also note that $b \geq -Q_{j,k}$ when $j \neq k$.

We now obtain (8.29). Let

$$s := s(l,k) = \frac{b}{1 - b U_{l,k}}. \tag{8.36}$$

By Lemma 5.4 it suffices to show that for each i, n in $\overline{\mathbb{N}}$.

$$\sum_{j=1}^{\infty} \left(U_{i,j} + s U_{i,k} U_{\ell,j} \right) \left(Q_{j,n} + b E(k,l)_{j,n} \right) = -I_{i,n}. \tag{8.37}$$

We first note that,

$$\sum_{j=1}^{\infty} U_{i,j} E(k,l)_{j,n} = \begin{cases} U_{i,k} & \text{when } n = l \\ 0 & \text{otherwise.} \end{cases} \tag{8.38}$$

We write this as

$$\sum_{j=1}^{\infty} U_{i,j} E(k,l)_{j,n} = U_{i,k} I_{l,n}. \tag{8.39}$$

It follows from this and Lemma 8.1 that,

$$\sum_{j=1}^{\infty} U_{i,j} \left(Q_{j,n} + b E(k,l)_{j,n} \right) = -I_{i,n} + b U_{i,k} I_{l,n}. \tag{8.40}$$

Using Lemma 8.1 again we also see that,

$$\sum_{j=1}^{\infty} U_{i,k} U_{\ell,j} Q_{j,n} = -U_{i,k} I_{l,n}, \tag{8.41}$$

and by (8.39)

$$\sum_{j=1}^{\infty} U_{i,k} U_{\ell,j} E(k,l)_{j,n} = U_{\ell,k} U_{i,k} I_{l,n}. \tag{8.42}$$

It follows from the last four equations that to get (8.37) we must have,

$$\left(b - s + sbU_{l,k}\right) U_{i,k} I_{l,n} = 0, \qquad \forall i, n \in \overline{\mathbb{N}}, \tag{8.43}$$

which follows from (8.36), since, $b - s + sbU_{l,k} = 0$. $\qquad\qquad\square$

Remark 8.2 Consider (8.29) with $k = l$. Then we can write

$$W_{i,j} = U_{i,j} + c_i c_j, \qquad \forall j, k \in \overline{\mathbb{N}}, \tag{8.44}$$

where $c = \{c_i\} \in \mathbf{Z}$ is a sequence of real numbers.
  If $k \neq l$, unless

$$U_{i,k} U_{l,j} = U_{j,k} U_{l,i}, \tag{8.45}$$

W is not symmetric. Furthermore, unless

$$U_{i,k} U_{l,j} = f_j, \qquad \forall i \in \overline{\mathbb{N}} \tag{8.46}$$

W does not have the form of (1.12). It has the form

$$W_{i,j} = U_{i,j} + c_i d_j, \qquad \forall j, k \in \overline{\mathbb{N}}. \tag{8.47}$$

where $d = \{d_j\} \in \mathbf{Z}$ and $d \neq c$. In these cases W is a new class of non-symmetric kernels for permanental processes.

Example 8.1 Consider the matrices Q and $\widehat{W}$ in (3.54) and (3.55) and create the Q matrix,

$$\tilde{Q} = Q + bE(1,2), \tag{8.48}$$

where

$$b = \frac{b'}{1 - r^2}, \tag{8.49}$$

so that this first row of $\tilde{Q}$ is

$$\frac{1}{1 - r^2}(-1, r + b', 0, 0, \ldots), \tag{8.50}$$

and all the other rows are unchanged. Since $r < 1$ there are values of b for which $\widetilde{Q}$ is a Q matrix.

Using (8.28) and (8.29) we see that the potential corresponding to $\widetilde{Q}$ is $\widetilde{W} = \{\widetilde{W}_{i,j}; i, j \in \overline{\mathbb{N}}\}$ where,

$$\widetilde{W}_{i,j} = \widehat{W}_{i,j} + \frac{b\widehat{W}_{i,1}\widehat{W}_{2,j}}{1 - b\widehat{W}_{1,2}} = \widehat{W}_{i,j} + \frac{b\widehat{W}_{i,1}\widehat{W}_{2,j}}{1 - br}. \tag{8.51}$$

In particular

$$\widetilde{W}_{1,1} = \widehat{W}_{1,1} + \frac{b\widehat{W}_{1,1}\widehat{W}_{2,1}}{1 - br} = 1 + \frac{br}{1 - br}, \tag{8.52}$$

and for $j \geq 2$,

$$\widetilde{W}_{1,j} = \widehat{W}_{1,j} + \frac{b\widehat{W}_{1,1}\widehat{W}_{2,j}}{1 - br} = r^{j-1} + \frac{br^{j-2}}{1 - br} \tag{8.53}$$

and

$$\widetilde{W}_{j,1} = \widehat{W}_{j,1} + \frac{b\widehat{W}_{j,1}\widehat{W}_{2,1}}{1 - br} = r^{j-1} + \frac{br^{j}}{1 - br},$$

which shows that $\widetilde{W}$ is not symmetric. For $j, k \geq 2$,

$$\widetilde{W}_{j,k} = r^{|j-k|} + \frac{br^{j+k-3}}{1 - br}, \tag{8.54}$$

so, for these values, $\widetilde{W}_{j,k} = \widetilde{W}_{k,j}$.

Note that for $\widetilde{Q}$ to be a Q matrix we must have $r + b' \leq 1$. Therefore, by (8.49), we must have $b \leq 1/(1 + r)$. Consequently we see that for $j \geq 2$, $\widetilde{W}_{j,1} < \widetilde{W}_{1,j} \leq \widetilde{W}_{1,1} = \widetilde{W}_{2,2}$, although obviously,

$$\lim_{j \to \infty} \widetilde{W}_{j,j} = 1. \tag{8.55}$$

Let $X_\alpha = \{X_\alpha(n), n \in \overline{\mathbb{N}}\}$ be an α-permanental sequence with kernel $\widetilde{W}$. It follows from Theorem 1.4 and (8.55) that for all $\alpha > 0$,

$$\limsup_{n \to \infty} \frac{X_\alpha(n)}{\log n} = 1, \quad a.s. \tag{8.56}$$

Bibliography

1. R.B. Bapat, Infinite divisibility of multivariate gamma distributions and M-matrices. Sankhya **51**, 73–78 (1989)
2. R. Blumenthal, R. Getoor, *Markov Processes and Potential Theory* (Academic Press, New York, 1968)
3. C. Dellacherie, S. Martinez, J. San Martin, *Inverse M-Matrices and Ultrametric Matrices*, Lecture Notes in Mathematics, vol. 2118 (Springer, NY, 2014)
4. E.B. Dynkin, Gaussian and non-Gaussian random fields associated with Markov processes. J. Fncl. Anal. **55**, 344–376 (1984)
5. N. Eisenbaum, H. Kaspi, On permanental processes. Stoch. Process. Appl. **119**, 1401–1415 (2009)
6. N. Eisenbaum, F. Maunoury, Existence conditions of permanental and multivariate negative binomial distributions. Ann. Probab. **45**, 4786–4820 (2017)
7. D. Freedman, *Markov Chains*, 2nd edn. (Springer, NY, 1983)
8. R.C. Griffiths, Characterizations of infinitely divisible multivariate gamma distributions. J. Multivar. Anal. **15**, 12–20 (1984)
9. V.A. Koval, The law of the iterated logarithm for Gaussian sequences and its applications. Theor. Probab. Math. Stat. **54**, 69–75 (1997)
10. M.B. Marcus, J. Rosen, Sample path properties of the local times of strongly symmetric Markov processes via Gaussian processes. Ann. Probab. **20**, 1603–1684 (1992)
11. M.B. Marcus, J. Rosen, *Markov Processes, Gaussian Processes and Local Times* (Cambridge University Press, New York, 2006)
12. M.B. Marcus, J. Rosen, A sufficient condition for the continuity of permanental processes with applications to local times of Markov processes. Ann. Probab. **41**(2), 671–698 (2013)
13. M.B. Marcus, J. Rosen, Conditions for permanental processes to be unbounded. Ann. Probab. **45**, 2059–2086 (2017)
14. M.B. Marcus, J. Rosen, Permanental random variables, M-matrices and α-permanents. *High Dimensional Probability VII, The Cargèse Volume, Progress in Probability*, vol. 71, pp. 363–379 (Birkhauser, Boston, 2016)
15. M.B. Marcus, J. Rosen, Sample path properties of permanental processes. EJP (58), 1–49 (2018)

© The Author(s), under exclusive license to Springer Nature Switzerland AG 2021   111
M. B. Marcus and J. Rosen, *Asymptotic Properties of Permanental Sequences*,
SpringerBriefs in Probability and Mathematical Statistics,
https://doi.org/10.1007/978-3-030-69485-2

16. M.B. Marcus, J. Rosen, Permanental sequences with kernels that are not equivalent to a symmetric matrix. *High Dimensional Probability VIII, The Oaxaca Volume*, pp. 305–320 (Birkhauser, Boston, 2019)

17. J.R. Norris, *Markov Chains* (Cambridge University Press, New York, 1999)

18. D. Revuz, M. Yor, *Continuous Martingales and Brownian Motion*, vol. 293, 3rd edn. (Springer, Berlin, 1999)

19. P. Revesz, *Random Walk in Random and Non-random Environments* (Cambridge University Press, New York, 1990)

20. H. Taylor, S. Karlin, *An Introduction to Stochastic Modeling*, 3rd edn. (Academic Press, NY, 1998)

21. D. Vere-Jones, Alpha-permanents. N. Z. J. Math. **26**, 125–149 (1997)

Index

Symbols
$U(l, n)^{-1}$, 5
L_t^x, 20
$M^{j,k}$, 5
$\mathcal{B}$, 10
$\mathcal{D}_{n,+}$, 4
$\mathbb{N}$, 4
ϕ, 5
b-transform, 35
c^*, 14
$\mathbf{1}_m$, 75

B
Birth and death processes
    symmetric, 19
    with emigration, 10, 33
    without emigration, 8, 19
Borel right process, 4, 89, 103
Brownian motion, 20
    killed, 57

C
Cemetery state, 20
Covariance, viii, 3, 58, 59, 61, 69, 95

E
Eisenbaum and Kaspi, 2, 4
Excessive function, 4, 21, 33, 40, 51
Exit time, 103

F
Fitzsimmons, 17

G
Gaussian autoregressive sequence
    first order, 11, 47, 58
    k-th order, 13, 61
Gaussian processes, viii, 3
Gaussian squares, 5

H
Holding time, 21

I
Increasing function, 9
Infinitely divisible, 28, 59
Inverse M matrix, 59, 87–89

J
Jump matrix, 20

K
Kernel, 4, 87, 88, 97
Koval's Theorem, 27, 38

L
Law of the iterated logarithm, 74
Lévy processes, 14, 100

© The Author(s), under exclusive license to Springer Nature Switzerland AG 2021   113
M. B. Marcus and J. Rosen, *Asymptotic Properties of Permanental Sequences*,
SpringerBriefs in Probability and Mathematical Statistics,
https://doi.org/10.1007/978-3-030-69485-2

Linear regressions
    first order, 82
    k-th order, 67
Local time, 20

M
Markov chain, 40, 48, 49, 57, 59, 62, 77
    uniform, 7, 104
Markov processes, viii, 3
M matrix, 77

O
O'Bryant, 17
Ornstein-Uhlenbeck process, 11

P
Permanental random variable, vii
Permanental sequence, 97
Population models, 13
Potential, viii, 34, 40, 48, 49, 58, 62, 66, 67,
    77, 78, 103

Potential density, 3
Potential function, 4, 21

Q
Q-matrix, 7, 8, 10, 13, 19, 33, 35, 56, 62, 63,
    66, 67, 78, 104

R
Regularly varying function, 54
Riesz decomposition theorem, 25

S
Stopping time, 78
Strictly concave function, 9
Symmetrizable, 4

T
Töeplitz matrix, 13, 63
Tridiagonal, 33

Printed in the United States
by Baker & Taylor Publisher Services